Covered Bridges

1. Beaver Creek Bridge
2. Smith's Bridge
3. Thompson's Bridge
4. Rockland Bridge
5. Rising Sun Bridge
6. Ridele's Bridge
7. & 8. Market Street Bridge, 1822 & 1839
9. Yorklyn Bridge
10. Ashland Bridge #118—still standing
11. Ashland Bridge #119
12. Mt. Cuba Road Bridge #120
13. Covered bridge washed out in 1938 flood
14. Red Clay Creek Bridge at Ashland
15. Mt. Cuba Road Bridge southeast of Ashland
16. Wooddale Bridge #137—still standing
17. Kiamensi Bridge
18. Bridge near Brandywine Springs on Hyde Run
19. White Clay Creek Bridge near Harmony
20. White Clay Creek Bridge near Ruthby
21. Paper Mill Bridge
22. White Clay Creek Bridge at Thompson's Station
23. Yeatman's Bridge
24. Summit or Buck Bridge
25. Westminster Bridge
26. Covered Bridge Farms Bridge

BRIDGES

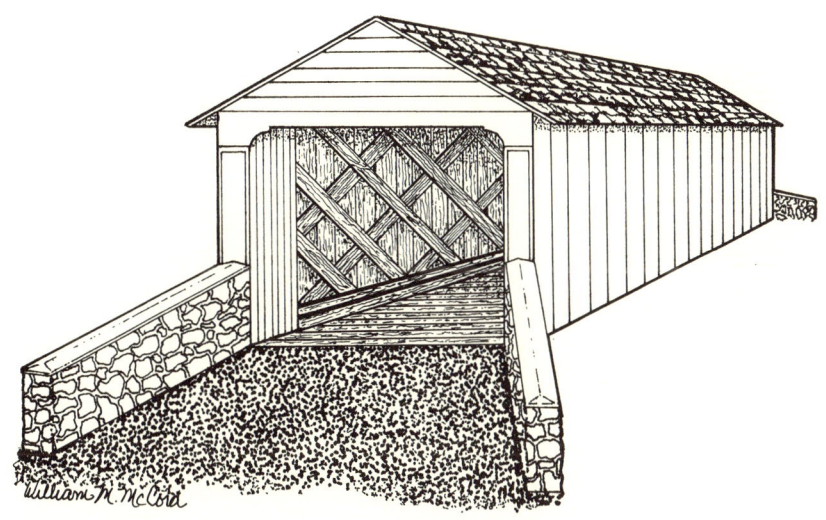

BRIDGES
Marjorie G. McNinch

Wilmington, Delaware

THE CEDAR TREE PRESS, INC.

PUBLISHED BY THE CEDAR TREE PRESS, INC.

First Edition

Published in a limited edition of 800. The first 25 copies reserved by the publisher. The remaining copies distributed to friends of The Cedar Tree Press.

This is copy number _____

Copyright 1995 by *The Cedar Tree Press, Inc.*
All rights reserved. No part of this book may be reproduced in any form without written permission from the publisher.
Manufactured in the *United States of America*.

Cover illustration by William M. McCord.

Marjorie G. McNinch

ACKNOWLEDGEMENTS

The author would like to thank the following individuals and organizations for their assistance in the preparation of this book: Special Collections at the University of Delaware Morris Library; the Historical Society of Delaware; The Delaware State Archives; the Hagley Museum and Library; the Wilmington Institute Free Library; Dennis O'Shea and his staff at the Delaware Department of Transportation; Richard C. Layton; Terry A. Bryan, D.M.D., Steven C. Gregory, Jr., James W. McNinch and Bruce D. Gasparo.

Marjorie G. McNinch

INTRODUCTION

This is the sixth book in our series on Delaware history and it is by far the most far reaching volume we have published. Our previous editions pinpointed small pockets of history: Rockford Tower, The Curtis Paper Mill, The King Street Market, P.S. duPont High School and Judy Johnson. Here we survey the entire northern Delaware county and discover that there once were at least 26 covered bridges within New Castle County.

Now considered a quaint reminder of bygone times, these bridges played an important part in the economic growth of the northern part of our state. The covering of bridges was also done for economic reasons, not to keep horses from being "spooked" because they could see over the side of the bridge.

The less the deck of a wooden bridge is exposed to the elements, the longer it will last. The person paying for the bridge, whether the county, the state or a private party, was willing to spend more to protect his bridge because this extended its economic life. Of course, today's bridges, made of concrete and steel, don't need this protection.

I remember and have driven across a few of the bridges described here, but only a few. I also remember the headlines when Smith's bridge was burned down one Halloween night. Even though I was a teenager at the time and prone to my own mischief night pranks, I felt a loss since I had travelled over that old bridge many times.

Steel and concrete, bricks and mortar long ago destined the covered bridge to go the way of the horse and buggy: still in limited use, but woefully out of date. There are only four covered bridges left in the county today. The Covered Bridge Farms bridge, the Westminster bridge, the Wooddale bridge, and the Ashland bridge. Of these, only the Ashland bridge is on a public road. The other three are on private property and not intended for general public use. The Ashland and Wooddale bridges are both listed in the National Register of Historic Places.

We have certainly come a long way since the day when the Summit covered bridge was built over the Chesapeake and Delaware Canal. That bridge was 225 feet long and was 90 feet above the canal. The canal itself cost $2,201,864 to build; the bridge much, much less.

This year we completed the most recent bridge crossing of that body of water. It has taken 3 years and 4 months to complete. It is 4,650 feet long, 127 feet wide, rises 138 feet above the water and cost $57,763,192 to build. And it isn't even covered.

Nicholas L. Cerchio III
Wilmington, November 1995

Marjorie G. McNinch

TABLE OF CONTENTS

Acknowledgement v

Introduction vii

Chapter 1: New Castle County's Covered Bridge History 3

Chapter 2: Building a Covered Wooden Bridge 11

Chapter 3: Brandywine Creek 17

Chapter 4: Red Clay Creek 41

Chapter 5: White Clay Creek 65

Chapter 6: One of a Kind Covered Bridges 79

Chapter 7: The Art of the Covered Bridge 85

Bibliography 89

Footnotes 95

Marjorie G. McNinch

ILLUSTRATIONS

Map of New Castle County Covered Bridges ....Endleaves

Market Street Bridge, 1860..4

Wooddale Covered Bridge 8

Interior of Smith's Bridge..12

Mt. Cuba Bridge..14

Map of Brandywine Hundred 18

Beaver Creek Covered Bridge...20

Truss Bridge that replaced Beaver Creek Bridge 20

Smith's Bridge before 1956 23

Rockland Mills, 1827..28

Rockland Bridge, 1932...28

Thompson's Bridge, 1930s 26

Side View of Thompson's Bridge 26

Rising Sun Bridge 30

Postcard of the Rising Sun Bridge 30

Rising Sun Bridge, 1995 32

Ridele's Banks near site of present
    Washington Street Bridge..34

View of Market Street Bridge from the 16th Street side................37

Covered Bridge at Market and 16th Streets, 1870........................39

Bridges

Map of Christiana Hundred 42

Yorklyn Bridge 45

Ashland Covered Bridge, 1995 46

Covered Bridge which once stood
  on the Ashland Farm of Henry B. du Pont 50

Mt. Cuba Road Bridge, 1930 52

Mt. Cuba Road Bridge, 1995 52

Wooddale Bridge 54

Kiamensi Bridge 58

Bridge near Brandywine Springs, 1920s 60

Map of Mill Creek and White Clay Creek Hundreds 64

Sketch of wagon being drawn through a covered bridge 66

White Clay Creek Bridge
  near Ruthby, 1965 68

White Clay Creek Bridge
  near Harmony, before 1925 68

White Clay Creek Bridge
  near Thompson's Station, 1920s 74

Covered Bridge over the Chesapeake and Delaware Canal 78

Covered Bridge at the entrance of
  Covered Bridge Farms, Newark, DE 80

Market Street Mills and Covered Bridge Illustration 83

Ashland Covered Bridge 87

Twenty men crossing a bridge,

Into a village,

Are twenty men crossing

  twenty bridges,

Into twenty villages,

Or one man

Crossing a single bridge into

  a village.

Metaphors of a Magnifico (1923)

BRIDGES

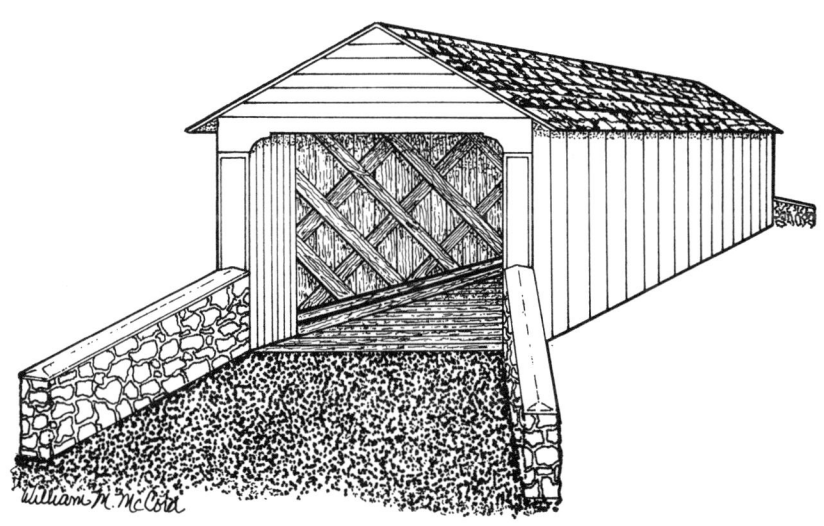

Marjorie G. McNinch

New Castle County's Covered Bridge History

New Castle County, Delaware has a covered bridge history. It is one of the county's best kept secrets. During the nineteenth century, New Castle County had approximately one hundred and twenty covered bridges crossing its creeks and canals, according to tradition. As the history of the county progressed into the twentieth century, thirty-five of these bridges remained. Now the county has only two standing historical covered bridges, the Ashland and Wooddale bridges over Red Clay Creek. The impressive attempt to preserve Smith's Bridge over the Brandywine Creek could not withstand the ravages of vandalism. The covered bridge standing at the entrance to Covered Bridge Farms outside of Newark was reproduced to give the development there its name. This reproduction also hints at the romance this county once had with the covered bridge. Almost four-fifths of the one hundred and twenty covered bridges of yesteryear are lost to history, since neither document nor photograph have been located to record their existence. The history of the twenty-five remaining covered bridges over the Brandywine Creek, the Red Clay Creek, the White Clay Creek, and the Chesapeake and Delaware Canal will reveal how they were connected with the commercial growth of the county, but did not survive its industrial growth. Nostalgia about the past helps to make it part of our history by releasing its secrets.

Settlements in New Castle County began in the city of Wilmington, Delaware, after Peter Minuit led the Swedish up the Christina River in 1638. Water transportation in the early 19th century was the main mode of travel and the main route to deliver goods, so manufacturers and farmers needed to be able to reach a port where their goods could be transported to major business centers. Following the Swedish came the Finns, the Dutch and the

Bridges

Market Street Bridge, 1860. Photograph by George M. Brinton. Courtesy of the Historical Society of Delaware.

English. Wilmington was settled by Quakers, who were encouraged to come to the area because of the Penn charter and because of the Delaware waterways. Families such as the Gilpins, Shipleys, Tatnalls, Canbys and Leas began enterprises along the Brandywine which gradually attracted more people and businesses to the area. By 1814 mills peppered the Brandywine; there were fourteen grist mills alone by then. Wilmington began forging ahead of New Castle, a significant port on the Delaware River, as an important port and business site. The growth of Philadelphia would eclipse this status by mid-century, but commerce and industrial growth continued in Wilmington as well as spreading west to Stanton and Newark. The farmers and merchants of Wilmington and the surrounding country transported their goods and raw materials overland to business and milling centers by wagons, as the public and private road network expanded. Timber bridges were originally constructed to cover the rivers and streams they crossed. Technology for the covered bridge from Connecticut was adopted in New Castle County as early as 1820. Timber bridges were covered to protect the planks from rotting as quickly as exposed planks. Thus the era of the covered bridge came to New Castle County, Delaware.[1]

The first covered bridge in New Castle County, for which documentation exists, is the Market Street Bridge over Brandywine Creek built in 1822 by Lewis Wernwag, who is said to have also built the Rockland covered bridge eleven years later. Wernwag is known for his covered bridge construction in Pennsylvania, particularly the "Colossus" over the Schuylkill River in 1812.[2] Three patented types of wooden bridges were used in the construction of the Delaware covered structures. Ithiel Town of Connecticut patented his truss form, the wooden lattice bridge, in January 1820. This bridge type was popular because the absence of any iron hardware, the "simplicity of the mechanical operations," and the ready availability of wood kept the expense down.[3] The next important covered bridge type was patented by William Howe, also of Connecticut, in 1840. This bridge type became the standard for wooden railroad bridges. The third type was patented by Theodore Burr, also of Connecticut, in 1804 and 1817, a trussed arch form.[4]

Covered bridge building in the nineteenth century was work done by contract, much like today, with the exception that the builder and the designer were usually the same man. Local saw mills cut the wood, and local blacksmiths provided the bolts and rods. The designer/builder would set the bridge frame on land, and "erect the bridge in place, ready for traffic." These bridge builders were part of a select few who had knowledge of bridge building, thus keeping competition at a minimum, allowing for control of "a certain territory."[5] In New Castle County the County Levy Court contracted for the construction of these bridges. Lewis Wernwag adopted the Burr arch truss in his construction of the North Market Street bridge and the Rockland bridge. The covered bridge at Rising Sun, constructed in 1833, and Smith's Bridge, in 1839, both over Brandywine Creek, were also Burr trussed arch bridges. The Town design was used in the Ashland, Mount Cuba, and Wooddale bridges, all over Red Clay Creek, and the Howe type was allegedly used in covered bridge construction on either Red or White Clay Creek.

Nineteenth century road networks throughout New Castle County affected both the population and commercial growth of the areas surrounding the three creeks, and in some ways caused unwanted competition between them. In Delaware, as well as in the other eastboard states, two types of roads existed. They were the post roads and the common roads. The post roads were the more heavily traveled and paralleled waterways; common roads traversed the entire state and connected less populated areas. By 1801 overland travel was faster and cheaper than waterway transport, and thus became the favored mode to travel or deliver goods. The expansion of the road network in New Castle County aided the development of commerce; the growth of trading centers in Wilmington, New Castle, Newport, Stanton and Christiana in turn boosted road expansion. The first improvement to the road system was the creation of private roads or turnpikes. These roads, mapped out to decrease travel time, were built and operated by private companies; the tolls collected paid for the road. These companies received charters by the Delaware State Assembly and they were required to complete the roads.[6]

By 1812 six turnpikes throughout New Castle County assured Wilmington its place as the county's commercial center, with Red Clay Creek giving the city heavy competition. In 1808 the Newport-Gap Turnpike connected Red Clay Creek with Lancaster, Pennsylvania. The New Castle-Frenchtown Turnpike was chartered to run from New Castle to the Delaware/Maryland border in 1809. In 1811 a turnpike was chartered between Wilmington and Newport, and in 1812 the Wilmington and Kennett Turnpike was chartered to run between Wilmington and the rich farmlands of the Delaware Valley. The Brandywine merchants petitioned for a better network to Philadelphia, so the the Wilmington and Great Valley Turnpike, completed by 1818, and the Wilmington-Philadelphia Turnpike, completed in 1823, assured Wilmington favored business status. A seventh turnpike, the New Castle-Wilmington Turnpike was also chartered in 1812. All of these turnpikes benefitted Wilmington, but three of them met the needs of the Red Clay Creek millers, although none were on the turnpikes. Few bridges are associated with these private roads. All the turnpikes were built in New Castle County because the economy and commerce supported them. Between 1825 and 1878 railroads eclipsed overland travel, causing road construction to decline.[7]

In early records of New Castle County bridges were referred to as wooden or timber, but not necessarily designated as covered. In the 1913 New Castle County Bridge book at the Delaware State Archives, eight bridges in Christiana Hundred are listed as wooden. Five of them have been documented as covered: Wooddale, Mount Cuba, East and West of Ashland Mills, and at Yorklyn Snuff Mills. The wooden bridge "near William J. Armstrong's Farm" eludes firm documentation as a covered bridge.

Bridges

Wooddale Covered Bridge showing exposed lattice truss work under cover. Photograph by Steven C. Gregory Jr., 1995.

Of the fourteen bridges listed in Mill Creek Hundred, four are known covered bridges: Kiamensi, Curtis Paper, Harmony Station and Thompson's Station. Inclusions on this list, such as "Wooden Bridge at Hockessin Chandler's Shop", are indicative of the widespread use of wood in smaller county bridges into the twentieth century, and are priceless in the overall historical documentation of bridge building in a geographical area. It is, however, frustrating to find no covered bridge designation. So be it. The county's documented covered bridges offer real connections with the past; they bridge time, you might say.[8]

Authenticated Covered Bridges in New Castle County Delaware

ON BRANDYWINE CREEK

    Beaver Creek
    Smith's Bridge
    Thompson's Bridge
    Rockland Bridge
    Rising Sun Bridge
    Ridele's Bank Bridge
    North Market Street

ON RED CLAY CREEK

    Ashland Bridge
    East of Ashland
    West of Ashland (same as the Mt. Cuba Road Bridge)
    DuPont Bridge in Ashland
    South of Ashland (or Red Clay Creek Bridge at Ashland)
    Mt. Cuba Road Bridge
    Mt. Cuba Bridge
    Wooddale
    Yorklyn Bridge

Bridges

    Kiamensi Bridge
    Bridge near Brandywine Springs

ON WHITE CLAY CREEK

    Harmony Bridge
    Ruthby Station Bridge
    Curtis Paper Mill Bridge
    Thompson's Station
    Yeatman's Bridge

OTHERS

    Authentic: Summit or 'Buck' Bridge over the Chesapeake and Delaware Canal
    New: Covered Bridge Farms, roadway into the development
    New: Westminister Bridge on Hyde Run

Marjorie G. McNinch

Building a Covered Wooden Bridge

New England master carpenter/mechanic Timothy Palmer built the first covered wooden bridge in the United States in 1805 across the Schuylkill River in Philadelphia at Market Street. It was a three span, 550 foot braced arch bridge constructed of heavy timbers. When it first opened, the bridge was not covered, whereupon Palmer insisted that a roof and siding be erected over it to keep the main frame timbers and arches dry to extend the bridge's life. Palmer called this type of bridge the "Permanent" bridge. Thus the history of covered bridges in the United States began, with an estimated count of over 10,000 covered bridges built between 1805 and 1885.[9]

Connecticut born Theodore Burr built the second covered wooden bridge in the United States over the Delaware River between Morrisville and Trenton, New Jersey, using his truss method of supporting and reinforcing bridges, which he patented in 1804. Burr's truss was a design incorporated in several of New Castle County's covered bridges, although Burr never built a bridge in Delaware. New Castle County's first known covered bridge was allegedly designed and built by Lewis Wernwag, a German immigrant who was a skilled mechanic and bridge builder, known, as Burr was known, for building long spans using heavy timber of white pine, a wood which would last. Wernwag built the county's first covered bridge over the Brandywine Creek at North Market Street in Wilmington in 1822.[10]

Before bridges over rivers and streams aided in making overland travel easier and faster, fords or ferries were often the predecessors. This was so along the Brandywine, particularly by the Lea flour mills on Market Street, and along Red Clay Creek around Newport and Stanton. With the development of public and private road networks running North/South and East/West within the county connecting rural areas with ports and business centers,

Interior of Smith's Bridge showing the arched truss bridge type patented by Theordore Burr in 1817. Photograph by Steven C. Gregory Jr., 1995.

the demand for bridges grew; farmers and traders were eager to deliver their goods, be they raw materials or manufactured wares, to market. Nineteenth century bridge builders used timber extensively, particularly for short spans, with masonry arch construction occasionally. The use of timber and temporary constructions were advisable in the early 1800s because labor for costly stone construction was lacking, plus load requirements and sizes of structures were uncertain. Timber was cheap and readily available. Timber bridge types in Delaware include: single span timber beam on pile bents, multiple span timber beam; covered wooden truss; timber beams on masonry abutments; and simple timber trusses on masonry abutments. Covered bridges were built only in New Castle County, not in Kent or Sussex Counties.[11]

Construction of covered bridges throughout the United States followed designs by New England bridge designers, such as Theordore Burr, Ithiel Town, William Howe, Col. Stephen H. Long, Peter Paddleford, and Timothy Palmer. Lewis Wernwag, mentioned earlier, was a reputed local bridge designer and builder. The covered bridge designers and builders who are responsible for supplying the demand for covered bridges in New Castle County are Lewis Wernwag, Theodore Burr, Ithiel Town and William Howe. Burr was a designer/builder, although he built no covered bridges in Delaware; Town and Howe, both from New England, were designers only, who eventually sold their rights to their design patents. To them New Castle County owes our covered bridge heritage.

Ithiel Town patented the wooden lattice truss bridge form, a form quickly adopted throughout New England and the Mid-Atlantic states. The fundamental ingredient of Town's bridge truss form is the triangle, "the only geometrical figure that cannot be distorted under load." Town's design "consisted essentially of top and bottom chords placed parallel with vertical posts located at the truss ends and with inclined web members forming a lattice work." The timbers used were of uniform size, usually 2 to 4 feet thick, and 10" to 12" wide, and were commonly referred to as plank. For longer spans, web members were placed closer together, and the chords were made in two sets, one placed above the other. The Town covered bridge could be assembled simply without bolts, rods or straps of iron; trunnels or treenails were used. Spruce was generally used, which tended to warp quicker than white pine, but lasted longer when properly covered. This lattice truss design was very popular because the planks were readily obtainable, and the cover could be assembled on land, and set in place over the timber beams. Often the lattice work was combined with the arch, which made a stronger bridge.[12]

Theodore Burr's covered bridge design consisted of an arch combined with the truss, "the arches springing from the face of the abutment below the bridge seat." His bridges were known for their longevity because he used heavy timbers of white pine. This type of bridge required more skilled labor, as well as less uniform timber sizes than the Town type, and was adapted more for longer covered bridge spans. "The posts were hewn, not only to make seats for the main diagonals, and for bearings on the arches, but were notched

Mt. Cuba Bridge showing the lattice truss bridge type patented by Ithiel Town in 1820. Courtesy of the Historical Society of Delaware.

at the lower chord intersections. They acted in tension and had to be framed accordingly."[13] Burr also incorporated the multiple kingpost truss, and modified the flooring to be level, not curved. In fact in 1817 Burr patented his arch truss design using the multiple kingpost truss. Although Burr's design was costlier and more complicated, his design was a favorite type, although less so than Town's.

William Howe patented a truss bridge design in 1840, which became very popular, and set the standard for wooden railroad bridges. In his design, which was improved in 1846, the arch beam was combined with the general truss so it cooperated with the other parts of the truss. His bridge could be completely stressed analyzed by the mathematical practice then in use. The chords and braces were made of timber, and the vertical web members were made of round iron.[14]

Lewis Wernwag, like Burr, used heavy arches in his bridges with flared multiple kingpost trusses on either side. His design and construction of the "Colossus" over the Schuylkill River at Fairmount, Philadelphia in 1812 is testament to this. "Wernwag's practice was to saw all his timbers through the heart, to detect unsound wood and permit good seasoning of the timbers. He used no timbers of a greater thickness than six inches, and separated all the sticks of his arches by cast washers, to allow free circulation of the air." The arch was the main carrier of loads. It was centered in the truss to hold the truss from distortion by heavy loads.[15]

The designs of these four bridge men were adopted in the covered bridges built in New Castle County with some alterations. Sizes, styles, and shapes of these bridges varied according to their location and use. "Some were 'single door' bridges; others were 'dual highways.' Some were closed in all around with only vents on the sides, like Thompson's Bridge."[16] The sidewalks of the North Market Street Bridge were outside the covering of the bridge. Single spans were built wide enough for a team of horses drawing a load of hay, and designed to handle a weight limit from three to eight tons. At the Lea Flour Mills on the Brandywine at North Market Street, conestoga wagons, referred to as 'prairie schooners', with teams of six, eight or twelve horses crossed the covered wooden bridge. With the advent of the automobile, some closed in bridges had square-like openings cut along the sides of the bridge so drivers could see oncoming traffic. Most of the county's bridges were constructed of spruce, white pine or oak, most often hand-hewn and well seasoned. The tighter the covering and the more seasoned the wood, the longer the bridge would remain in service.[17]

Covered bridges served as vital links between communities and within communities. Since the covered bridge usually offered the largest covered area in some communities, the bridge became used for political rallies, revival meetings, weddings, a place for courting, and in case bad weather arose, shelter. Because the covered bridge offered an intimate place for romancing, it was sometimes labeled the "kissing bridge." The covered bridge, as has been noted, was of much more importance. The covered bridge helped a community to grow economically by providing a strategic link to commercial centers, and as areas expanded geographically more bridges were built. The service of the covered bridges discussed on the following pages will offer testimonials to the craftsmanship of the early covered bridge designer/builders.[18]

Marjorie G. McNinch

Brandywine Creek

Descriptions of Brandywine Creek all agree on its picturesque qualities. Elizabeth Montgomery in her *Reminiscences* of 1851 refers to the Brandywine as "a crooked creek," "a whimsical stream," and "a romantic stream." William Cullen Bryant in his *Picturesque America* of 1872 says of the Brandywine that "no other [river] unites the beauty of wooded heights and tumbling waterfalls with structures of art (mills) that give rare charm and even quaintness to the picture." Wilmer MacElree remarks in his 1906 description how the creek flows from its source in Honeybrook, Pennsylvania to tidewater in Wilmington, Delaware with "graceful sinuosity." The beauty of the Brandywine Valley as seen by these authors is reflected in the backdrop of the milling industry gracing its shores during the 19th and early 20th centuries. Bryant is convinced of the blending of these mills into the Brandywine's landscape. "Too often labor mars the landscape, but the mill seems to partake of the spirit of its surroundings, to gain a charm from woods and waters, and to give one. This is peculiarly true of the factories along the Brandywine." The rapid descent of the Brandywine from Welsh Mountains through Wilmington furnished the power to the milling concerns established along its banks from the Pennsylvania/Delaware boundary to the Brandywine's convergence with the Delaware River. Indeed it was this gradual one thousand foot drop which drew settlers to Wilmington to erect their businesses on this creek. The Brandywine, named in honor of a Dutch ship filled with brandy and wine which shipwrecked in 1665 in the river, is the site where eight (two at North Market Street) of New Castle County's covered bridges spanned its waters to aid the enterprises needing access to markets.[19]

As one travelled downstream on the Brandywine Creek from the boundary with Pennsylvania in the mid 19th century, the first area entered would be Beaver Valley. Here stood two covered bridges: Beaver Creek serving possibly the Sunnydale Paper Mill;

Bridges

Brandywine Hundred. Daniel G. Beers, <u>Atlas of the State of Delaware</u> (Philadelphia, PA: Pomeroy & Beers, 1868).

and Smith's Bridge named for and serving Isaac Smith and Joseph Brinton, who operated a flour and grist mill there. The next covered span to be encountered is Thompson's Bridge across which "loads of hay and grain drawn by horse teams with slow and measured tread" were seen. Constable Davey Wilson, who lived above the bridge, most likely watched the constant traffic. Below Thompson's one finds the village of Rockland where a covered bridge spanned the creek to allow the owners and the workers of the Rockland Manufacturing Company to transport their cotton and woolen goods. Not far from here was the DuPont Gunpowder Works, taking advantage of the 125 foot drop of the creek to power its rolling mills. Henry Clay Village was just beyond the works, and it is here where the powder workers and their families lived. Rising Sun Bridge, also known as New Bridge, was a covered wooden span connecting the community and aiding in the transport of goods to Wilmington and Philadelphia. The covered bridge at Ridele's Bank near the old Barley Mill in the City of Wilmington proper was the beginning of a stretch of flour, grist, paper and saw mills, known as the Brandywine Mills, powered by the Brandywine's rapid descent to tidewater at this point. The Leas, Tatnall, Canbys, and Shipleys all had enterprises here. In 1814 fourteen grist mills alone lined the banks of the Creek. The North Market Street juncture with the Creek was the site of several bridges, two of them covered. The channel, narrow and rocky, opens up as it passes the Lea Mills and flows towards the Delaware River.

The following bridges across the Brandywine Creek are discussed in this chapter.

Beaver Creek Bridge (also known as Beaver Valley Creek): built, date unknown.

Smith's Bridge. County Bridge #9 on Road 221: built in 1839, restored and reinforced in 1955, burned down in 1961 and not replaced; Burr arch truss; 159 feet long and 16 feet wide.

Thompson's Bridge. County Bridge #3 on Road 225: built

Bridges

*View of Beaver Creek covered bridge, 1965.
Courtesy of the Historical Society of Delaware.*

View of truss bridge that replaced it. Courtesy of the Historical Society of Delaware.

in 1854, replaced in 1935; Burr arch truss; 171 feet long and 16 feet wide.

Rockland Bridge. County Bridge #2 on Road 232: built in 1833, replaced in 1934; Burr arch truss built by Lewis Wernwag; 108 feet 6 inches long, 14 feet 6 inches wide.

Rising Sun Bridge. County Bridge #1 on Road 267A: built in 1833, replaced in 1928; Burr arch truss; 124 feet long, 22 feet wide.

Ridele's Bank (near site of the Washington Street Bridge): built, date unknown; Burr arch truss.

North Market Street Bridge. State Bridge #575 on State Route 13: 1st covered bridge built across Brandywine in 1822, 2nd covered bridge built in 1839, replaced in 1887; Burr arch truss built by Lewis Wernwag.[20]

Beaver Creek Bridge

Very little is known about this bridge, which covered a tributary of Brandywine Creek near the border of Pennsylvania and Delaware. Delaware State Highway Department photographs of the covered bridge and its replacement (a 1965 view) is clear documentation that it existed, but exactly where and for how long has proven elusive. In a letter written by covered bridge author Richard Sanders Allen to David K. Witheford, Planning Engineer of the Delaware State Highway Dept. in 1959, Allen writes, "The picture of Beaver Creek Bridge shows a low wooden truss bridge in which the trusses only are covered, and there is no roof. This was a transition type of wooden bridge and today is actually rarer than the full-sized roofed and weather-boarded type."[21]

Beaver Valley Creek Bridge, which may have replaced the covered wooden bridge known as Beaver Creek, is cited as being an early pony truss bridge built in 1890 by the Edge Moor Bridge Company in a 1982 historical bridge survey of Delaware. The modern photograph looks similar to a pony truss bridge. A second bridge of reinforced concrete was built in 1920 in this same location. Downstream from Smith's Bridge is a Brandywine tributary known as Beaver Run. The Sunnydale Paper Mill was built there in 1811, so perhaps Beaver Creek served Sunnydale. The Department of Transportation archives records bridge #8 as being on County Road 221 named Beaver Valley Creek, which would place the bridge on the Brandywine near the border with Pennsylvania. Conclusions here are difficult to draw. The photograph definitely documents a Beaver Creek covered wooden bridge, so that much is certain. Its location is also fairly certain. The dates of construction and dismantling, at this writing are unknown. The industry it served was grist milling. By the mid-1600s Swedish grist mills were built on both Harvey's Run and nearby Beaver Creek, both tributaries of the Brandywine.[22]

Smith's Bridge
County Bridge #9 on Road 221

"Above Smith's Bridge the stream flows along quietly, with overhanging vines on both sides, ministering alike to weary bodies and tired minds," so says a traveler walking along the Brandywine in 1912. Smith's Bridge is the next covered bridge near Beaver Creek to which we come. Smith's Bridge is named for Isaac Smith and Joseph Brinton, who operated a flour and grist mill there in the early 19th century. The first bridge at the mill site was built in 1816 on piers, but it washed away in the freshet of 1822. It was most likely a timber bridge, only not covered. The bridge was rebuilt and was washed away in 1828. The covered wooden bridge revered by Delawareans and covered bridge lovers was built in 1839 at a cost of $5,446. It was a one span Burr arch truss bridge, 159 feet long and 16 feet wide built on high stone abutments with "rusticated wingwalls of granite," and with steep approaches to keep the bridge beyond the reach of future floods. Isaac Smith operated a large farm, as well as the mill, and he sent produce to Philadelphia by wagon every week. His son Edward Smith

Smith's Bridge before restoration in 1956. Photograph by Nathaniel Rand. Courtesy of the Hagley Museum and Library.

continued operation of the mill, and was succeeded by Charles and James Twaddell, who ran it for fifty years. William T. Talley took it over, and the mill was last operated by his son, John T. Talley in 1900. "The power for the mill was obtained by means of a long race without a dam.[23]

After 116 years of service, this last covered bridge to span the Brandywine was damaged by an "exceptionally heavy truck" weighing more than the three ton limit; the bridge thus became unsafe for motorized traffic in 1954. A plea for its preservation was made by the state's chief engineer, Col. William A. McWilliams, and the State Highway Commission approved the reinforcement and overhaul in December 1954. Col. McWilliams was impressed with the excellent condition the housing was in. The "bow arches are in as good condition as the day they were put in place by the unknown artisans who built the bridge." In 1955 Smith's Bridge's roof was replaced, the floor replaced with steel flooring, the sides were repaired, the bridge was repainted barn red and the trim white, and "two piers were constructed with reinforced concrete faced with granite to blend with the original masonry." The cost was $43,125, and the contract was handled by Conn Welding and Machine Company of New Castle, Pennsylvania. Some of the timbers replaced were heavy cross beams 16" x 8", beams which had to have been cut to order by a saw mill, since they were much larger "than any wooden beams" used in modern construction. The effort to reconstruct the bridge without destroying its original outward appearance was admirably embraced by the construction team, including a plank floor placed over the steel. A three span bridge resulted with a weight limit raised to fifteen tons.[24]

During the year 1960 Irving Warner of the Warner Company wrote to his friends Henry F. duPont, founder of the Winterthur Museum, Maurice duPont Lee, President of the Wilmington Park Commission, and Richard A. Haber and Joe S. Robinson, Chief Engineer and Assistant Chief Engineer of the State Highway Department about preserving Smith's Bridge by moving it to a safe haven. Perhaps Mr. Warner had an inkling of what was coming.

The following year, 1961, just six years after the reconstruction, Smith's Bridge went up in smoke, torched by Halloween tricksters. The bridge was rebuilt to accommodate a cover, but no cover was ever placed over it perhaps because of monetary constraints, perhaps because of lack of public interest, but almost assuredly because of an underlying fear that the bridge would burn again.[25]

Smith's Bridge is now lost to history by its sad end. The 122 year old bridge was certainly a testament to the longevity of Theordore Burr's arch truss design and to the skill of the carpenters who built the bridge. Smith's will not be forgotten, however.

Thompson's Bridge
County Bridge #3 on Road 225

Thompson's Bridge is located in what the Delaware Department of Transportation calls Woodlawn Park. Thompson's Bridge was a Burr arch truss wooden bridge, one span 171 feet long and 16 feet wide built in 1854 and replaced eighty years later. The unique feature of this bridge was that it was closed in all around with only vents on the sides. Thompson's Bridge was on the east-west public route between Naamen's Creek and the Red Clay Creek, and thus served as a trade route. Residents in the area during the 19th century include Constable Davey Wilson and Betsy Kennedy, whose log house stood on what became known as Kennedy Hill.[26]

Newspaper reporter the late William P. Frank in a February 1934 article announced that Thompson's Bridge had served its time and would be replaced. During its eighty years the bridge was repaired once in 1919. The contract went to E. B. Hollingsworth, who used "first class white oak" to repair the planking, floor beams and stringers. By April 1935 a new concrete bridge was opened to the motor traffic which doomed the wooden structure to demolition. Thompson's Bridge was one of the most striking examples of a covered bridge because of its length, its small vent windows and its pastoral setting.[27]

Bridges

Thompson's Bridge, 1930s. Frank Zebley scrapbook. Courtesy of the Historical Society of Delaware.

Thompson's Bridge from the side. Courtesy of the Hagley Museum and Library.

Rockland Bridge
County Bridge #2 on Road 232

Rockland Bridge was a one span covered wooden bridge which crossed the Brandywine just below the Rockland Dam, which powered the William Young paper mills and later the Rockland Manufacturing Company cotton and woolen mills. The Rockland Bridge was of the Burr arch truss type, and was allegedly built by designer/builder Lewis Wernwag in 1833. It was shorter than Thompson's and Smith's, being 108 feet long and 14 feet 6 inches wide. The lumber used in its construction was mostly hand hewn white pine, "fitted together with the utmost care and precision." The maximum load it was designed to handle was six tons, although in the first half of the 19th century a hay wagon and team of horses was probably the heaviest load it had to carry.[28]

The village of Rockland was originally known as Kirk's Ford. When William Young, a Philadelphia printer, came to the area in 1793, he established paper mills on the east bank south of the dam. Caleb Kirk, a miller on the west bank, and Young entered into an agreement on the upkeep of the dam. In 1814 Young's paper mill burned and he rebuilt it as a woolen mill under the name of the Rockland Manufacturing Company. The first bridge to span the Brandywine at this point was built in 1818, and was known as Young's Bridge. The village became known as Youngstown. Young converted his woolen mill to a cotton mill in 1822. He died in 1829 and the land and mills were sold at sheriff's sale in 1854 to Augustus E. Jessup and Henry duPont. The Young mill was torn down in 1890.[29]

Milling continued at the site well into the 20th century with the covered wooden bridge at Rockland continually serving the community and business center. A traveller into the Rockland area early in the 20th century remarked that Rockland Village was appropriately named. One saw stone fences enclosing stone houses. Even the trees clung "to some congenial rock." One hundred years after its construction, the Rockland covered bridge

Bridges

Sketch by Eleuthera (du Pont) Smith (1806-1876) of the Rockland mills showing the covered bridge, 1827.

Rockland Bridge, 1932. Courtesy of the Hagley Museum and Library.

was dismantled. County engineers found that "both ends of its braced arches decayed to a fraction of an inch," and was in danger of collapse due to the increased load crossing the bridge. The late Bill Frank reported that "a bit of these decayed arches" could be seen in the County Engineer's Office in 1934. The county engineers credited the longevity of the span to the "well seasoned timber" that kept the Rockland Bridge "alive". When Rockland was razed, old planking was ordered to be "carefully removed" and all sound lumber salvaged. On May 17, 1934 a news article announced the opening of the new concrete bridge at Rockland. The bridge was constructed by Charles H. Dunleavy of Coatesville, Pennsylvania for $39,250. The top deck type allowed for a clear view of the vista down the Brandywine, a feature the covered bridge did not offer. Then again the Rockland covered bridge was a part of the vista.[30]

Rising Sun Bridge
County Bridge #1 on Road 267A

After the Brandywine passes under the Rockland Bridge it begins a descent downhill on its way to tidewater in the City of Wilmington. The 125 foot drop occurs as the creek passes the DuPont Gunpowder Works allowing 60,000 tons of water to crash over the four dams constructed to direct water to power the roll mills. "The original Rising Sun Bridge was an impressive timber covered bridge comprised of a Burr trussed arch spanning 124 feet and resting on massive granite abutments." The bridge was built in 1833 of the same design as Smith's, Thompson's and Rockland, and in the same year as Rockland. The duPont family and the gunpowder establishment had only been at this location for twenty-one years by then. The powder mills were constructed of Brandywine granite from a quarry on site. This same granite was used in the Rising Sun Bridge abutments. The bridge was a double span type, and was twenty-two feet wide. A sign posted over the bridge read "Walk your horses over the bridge." Wagons loaded

Bridges

Rising Sun Bridge showing the portals of the two lane span. Frank Zebley scrapbook. Courtesy of the Historical Society of Delaware.

Postcard of the Rising Sun Bridge 1915. Courtesy of the Historical Society of Delaware.

with hay were sometime piled too high, and hay caught on the support beams over the entrance way of the covered bridge.[31]

The Rising Sun Bridge was demolished in 1927 and a new steel "single span riveted Pratt through truss bridge" was built in its place, opening in 1928. The firm of Harrington, Howard & Ash of Kansas City and New York served as consulting engineers for construction of the new bridge under the auspices of the New Castle County Levy Court. This steel truss bridge utilized the massive granite abutments of the covered bridge, and spans 127 feet. This bridge is in itself a rarity. It is only one of three Pratt through truss bridges still operating in New Castle County which date back into the 1920s. A granite masonry arch which originally spanned a tail race during the life of the covered bridge remains part of this steel truss structure.[32]

The Rising Sun covered bridge served the little hamlets near the DuPont works and the Riddle's and Joseph Bancroft's textile mills downstream. These hamlets went by such names as DuPont's Banks, Rising Sun, Henry Clay Village, Rokeby, Breck's Lane, Newbridge, Walker's Banks, Free Park, Chicken Alley and Louviers. The rocky banks and steep inclines caused a magazine writer in 1895 to refer to the area as "this topsy-turvy village without streets." The banks of row houses and some stone houses were built by the DuPont Company for their employees. The adoption of 'banks' into a hamlet name comes from these banks of row houses. The employees of the powder mills and textile mills who lived in these hamlets were the chief users of this bridge. The bridge was not a part of an important through road; it was strictly a local bridge. Rising Sun Bridge provided a community gathering place, particularly as the 19th century wore on, and workers' hours were regulated so they had more leisure time. Youths enjoyed "driving a wagon through a covered bridge." The bridge became a "nightly gathering place for many of the young men of the neighborhood, a low wall on the southern approach serving as a seat." After the turn of the century, there was a decline in housing and employment. The DuPont mills closed in 1921, and the

Bridges

Rising Sun Bridge, 1995. This Pratt through truss bridge replaced the Rising Sun covered bridge in 1928. Photograph by Steven C. Gregory Jr., 1995.

DuPont Experimental Station started expanding on the site of the DuPont Country Club on the northeast side of the bridge. The People's Railway Company connected Rising Sun with other small communities, and the advent of the car and better housing elsewhere took people away from the area. The covered bridge, which for almost one hundred years held a community together, and offered a splendid view of the Brandywine Creek in both directions, met its demise. A traveller to Rising Sun in the early 1900s wondered about the origin of the name Rising Sun. Then he saw the sunrise, which made him exclaim that the Brandywine Creek was made "beautiful as the River of God."[33]

Ridele's Bank Bridge
City of Wilmington

Very little is known of this covered bridge except, like the Beaver Creek bridge, it was known to have existed. Ridele's Bank covered bridge is conjectured to have spanned the Brandywine Creek northwest of Wilmington on or near the site occupied by the Washington Street Memorial Bridge. It appears to be a Burr arched truss bridge, but the year it was built is unknown. Author William Cullen Bryant in his book *Picturesque America* of 1871 refers to Ridele's (most likely James Riddle's) cotton mills as "a very little way up the stream" in a picturesque setting, and that the covered bridge crossed there near a revolutionary grist mill. Bryant also shows an illustration of the bridge next to the grist mills. Grist mills lined the Brandywine banks in the early 1800s (fourteen in 1814) so it is possible that the bridge crossed next to a grist mill near Ridele's Bank. Author Richard Sanders Allen, who possibly got his information and picture from Bryant's book, records on his list of Delaware covered bridges that Ridele's covered bridge had squared portals and was near a revolutionary grist mill. It is true that during the Revolutionary War Joseph Tatnall, who operated a flour and grist mill where a North Market Street span crossed over, ground corn for use of Washington's army at Valley Forge. Bryant

Bridges

Ridele's Banks near site of present Washington Street Bridge. William Cullen Bryant, editor, <u>Picturesque America</u> (New York: D. Appleton & Co.), Vol. 1, page 229.

contends that the grist mill in his picture with the covered bridge is Tatnall's mill.[34]

Thus it is probable that the covered bridge named Ridele's Bank actually crossed the Brandywine Creek somewhere between Rising Sun and North Market Street. Since Bryant refers to it as Ridele's, it would be logical that the bridge crossed closer to the Riddle and Joseph Bancroft textile mills, a site nearer to the Augustine Bridge, not the Washington Bridge.

North Market Street Bridge
State Bridge #575 over Brandywine Creek

The first covered wooden bridge over Brandywine Creek was State Bridge #575 built in 1822. It was a double span timber Burr arched truss bridge, 145 feet long and 28 feet wide costing $7558.23. Before 1762 the only means to cross the creek was by ferry or ford. That year the Delaware General Assembly authorized the construction of a timber bridge, on stone abutments, and it opened in 1764. By 1806 this structure had undergone numerous repairs, so the public petitioned for construction of a stone arch bridge, forming a company with a capital of $20,000 to do so. Opposition from the area millers and merchants was so strong that the Levy Court had a chain link suspension bridge built for $4,000 instead. A flood washed this bridge away in 1822, so out of expediency New Castle County's first covered bridge over the Brandywine was constructed. This heavy timber bridge was itself washed away by flood in 1839. Lewis Wernwag, eminent bridge builder in the mid-Atlantic, built the second covered bridge at the site. The arches of this bridge were particularly heavy, and the portals ornate. The sidewalks of this bridge were covered overhead, but the side coverings were open. This Wernwag bridge was dismantled in a little less than fifty years, and was replaced in 1887 with a wrought iron and steel Pratt through truss, the same type of bridge which replaced the Rising Sun covered bridge. This Pratt truss bridge was replaced in 1928, whereas the one at Rising Sun is still in use.[35]

Where the Brandywine flows toward the Delaware River its banks become low and flat, and the river bed wider and deeper. Small boats, such as shallops and sloops could sail up as far as this juncture, so it is here where the line of mills started lining up stream. A ferry operated near where French Street reaches the creek until a timber bridge opened just north of there in 1764 at North Market Street. These grain boats bringing wheat raised on farms in the neighborhood, lower Delaware, nearby Maryland and New Jersey, operated under one large sail, but sometimes a tug had to pull them up to the narrowing juncture. This ground wheat or corn would either be taken home or sold to the miller.[36]

The millers began moving into this area of Wilmington by the mid-1700s. The cluster mills which were established at this juncture became known as the Brandywine Mills. The location was ideal because it was at the head of navigation of a powerful stream passing through the heart of the coastal wheat belt. Oliver Canby moved from Bucks County, Pennsylvania in 1741 and built the first flour mill about 200 yards above the present Market Street bridge on the south side. Thomas Shipley bought land along the Brandywine between the terminations of French and Market Streets in 1762. He built the first mill below the bridge on the south side, and soon built two others near it. Shipley's first mill was later bought by the City Water Department to use it as a pumping station. Joseph Tatnall, whose home was in Brandywine Village, inherited Shipley's mills and is said to be the only miller bold enough to grind corn for the Continental Army. Tatnall's son-in-law Thomas Lea, built the first flour mill on the north side just below the bridge. Another Tatnall son-in-law, James Price, built a mill on the south side below the bridge. By 1851 thirteen mills grinding grain operated below Market Street Bridge. Joseph Tatnall bought water rights on the north side of the Creek and sold them to William Lea, who operated extensive mills there through the 19th century. Superfine Lane off Race Street in Wilmington is the site of Lea's mills.[37]

What made this area a center for commerce was its location near the mouth of the Delaware River, its location on the north-south route between Philadelphia, and the power of the Brandywine Creek. Six dams are located between the end of DuPont's powder works and the tidewater end of the creek. Two dams supplied power to Joseph Bancroft's mills at Rockford and Kentmere. There is the Jessup and Moore dam below Bancrofts and a dam just below the Baltimore and Ohio Railroad Bridge. The next dam, at the foot of Adams Street, was known as the Barley Mill dam. The dam was destroyed in the freshet which destroyed the Market Street chain bridge; the dam was never rebuilt and the mill disintegrated. The last dam is at the foot of West Street. It furnished water to the north and south short races, powering mills below the bridge. One can readily see the importance of a substantial bridge at Market Street.[38]

Marjorie G. McNinch

Market Street Bridge from the 16th Street side looking across to Brandywine Village, c. 1887. Courtesy of the Historical Society of Delaware.

Brandywine Village on the north side of the bridge was not a part of the city, but of Brandywine Hundred, when the second covered bridge was constructed in 1839. A news reporter related that "The bridge was a sort of playground for 'Brandywine Village' children." After the village became a part of Wilmington, a Wilmington policeman looked out for the children who played there, and did not let them stray into town. It was also a gathering place for the "young men of the village" as well. Frank Zebley recounts a story about the second covered bridge before it was demolished in 1887 in his *Along the Brandywine*. A city resident, Amor H. Harvey of 1801 Market Street died at the time the bridge was slated to go. "The flooring of the old bridge was left in place until the funeral procession had crossed it on its way to Old Swedes Cemetery." The covered bridges over the Brandywine Creek at North Market Street served their constituents well.[39]

The last covered wooden bridge built to span the Brandywine Creek was Smith's Bridge in 1961. The covered bridges which spanned the creek from 1822 to 1961 remained standing from forty years to well over one hundred years, the 1822 North Market Street span having the shortest span of seventeen years due to a flood. The mills clustered around them originally were flour, grist, saw and paper mills. Gunpowder works and textile mills became dominant as 1900 approached. The bridges fostered commerce between the port of Philadelphia, Baltimore and the rest of New Castle County assuring Wilmington importance as a business center, thus expanding its growth and economy. Of the eight spans (two of them being at North Market Street), six were constructed of the arch truss design by Theodore Burr, a bridge of this design known for its longevity. Two were said to be built by designer/builder Lewis Wernwag, who used hand hewn heavy timbers of white pine in his bridge constructions. The Brandywine Creek alone sported covered wooden bridges to rival those of New England. The covered wooden bridges over Red Clay Creek, however, tell a parallel history, an impressive documentation of commercial growth along the Brandywine Creek's sister creek.

Portals of the double span covered bridge at Market and 16th Streets showing Phillips Mill (later the Wilmington Water Works Building site), 1870. Courtesy of the Hagley Museum and Library.

Marjorie G. McNinch

Red Clay Creek

Red Clay Creek is the smallest of the Delaware waterways flowing to the Christina River. It is only thirteen miles from its source in East Marlboro Township in Chester County, Pennsylvania to Stanton, Delaware where it joins the White Clay Creek on its passage to the Christina. For over three hundred years Red Clay Creek has driven a variety of mills, the first ones being sawmills built at Greenbank in 1677 and at Stanton in 1679. Both were converted to grist mills, and Greenbank still operates. Early reporter Hezekiah Niles described Red Clay Creek in 1815 as "a lively stream passing through a hilly country, abounding in springs and falls of water...and [giving] power to many establishments for various purposes." Grist mills and saw mills dominated the creek up to 1812 because they supported an agricultural economy, not a manufacturing one. The War of 1812 spurred the growth of textile, paper, iron, snuff, spice and plaster milling along its shores, and by the Civil War these mills, including grist, flour and sawmills, were thriving. By 1814 seven grist mills, six saw mills, two cotton mills and one snuff mill were operating on Red Clay Creek. Villages which grew clustered about these mills included Ashland, Wooddale, Marshallton, Mount Cuba, Yorklyn and Stanton. Niles again reported that Red Clay Creek's "banks are high but have many gentle, lawn-like slopes to the edge of the pure, swift-running stream," but enough ruggedness to make it picturesque. What economic developments occurred in Wilmington as well as in the United States were mirrored among the Red Clay Creek industries. The covered wooden bridges serving these industries were not as substantial as those crossing the Brandywine. Indeed the Red Clay Creek was not as powerful because it did not descend as rapidly as the Brandywine, nor was it as long. The Red Clay Creek milling communities built their bridges out of expediency, and because they were inexpensive to build, but not for longevity.

Bridges

Christiana Hundred. Daniel G. Beers, <u>Atlas of the State of Delaware</u> (Philadelphia, A: Pomeroy & Beers, 1868).

It is along Red Clay Creek, however, that New Castle County's two remaining covered wooden bridges stand.[40]

The majority of these bridges were built following the lattice truss design patented by Ithiel Town in 1820. These bridges were cheap to construct because wood was plentiful, the planking was uniform, and no iron rods or straps were needed to hold the truss together. The lattice truss cover could be constructed on land, and immediately placed over the timber span. As the Red Clay Creek crosses over the Pennsylvania/Delaware border, the first covered bridge one came to was at Yorklyn. John Garrett Jr.'s snuff mill began operation here in 1782, a large operation lasting into the 1950s. A Garrett family paper mill, the first on Red Clay Creek, was also established at this site in 1804, and was converted into a cotton mill, the Auburn Mill in 1813. The Auburn mill was run by the Pusey family from 1813-1880. Ultimately the Marshall family gained possession of the works and built a new paper mill there, which was kept running into the 20th century as well. Both the Garrett and Marshall families initially operated grist mills along Red Clay Creek. The next village where the creek was spanned by covered bridges is Ashland. Four of Red Clay Creek's covered bridges stood at or near Ashland, and one of them still does. It was at Ashland Station that a grist mill began operation in 1715 by the William Gregg family, and was operated for forty years by A. and J. D. Sharpless, who were succeeded by George W. Pusey in 1895. George W. Pusey was still operating the mill in the 1930s as a flour mill.[41]

Two covered bridges spanned Red Clay Creek at Mount Cuba, and one at Wooddale. A grist mill and saw mill were operating at this site by 1814. Wooddale is the site of the Marshall family's second paper mill and the Delaware Iron Works. The paper mill was built on the site of the Delaware Iron Works by the Marshall family in 1891, and operated into the 20th century. The Delaware Iron Works was on the site of the Gilpin slitting mill. Alan Wood and Co. steel works of Conshohocken, Pennsylvania leased, then purchased the ironworks in 1844. Two successful textile factories on Red Clay Creek were near covered bridges. The Kiamensi Woolen Factory, located near Stanton, began the manufacture of woolen goods in 1809 and was still in operation in 1914. The other textile mill was the Henry Clark operation on a tributary of Red Clay Creek, Hyde Run, running from 1837 into the 1890s.[42] The industrial development of the area occurred because of the economic growth of nearby Wilmington, advances in technology in the textile and ironworking trades, expansion of the road networks, and the age of the railroad. Red Clay Creek provided its waters to power the many mills established along its shores. The covered bridges aided in this growth, as they did in Wilmington. These bridges are discussed in this chapter.

Yorklyn Bridge. County Bridge #112 on Road 257: built 1863, replaced in 1929; Town lattice truss.

Ashland. County Bridge #118 on Road 258: built ca. 1840, still standing; Town lattice truss.

Ashland. County Bridge #119 on Road 261; south of Ashland, known as the "duPont Bridge" being on the property of Henry B. du Pont; it was purposely burned down ca. 1964.

Ashland. County Bridge #120 on Road 261, Mt. Cuba Road west of Ashland; replaced in 1922; Town lattice truss.

Ashland. On Road 261 east of Ashland; washed out in flood in 1938; Town lattice truss.

Ashland. Listed as Red Clay Creek Bridge at Ashland, New Castle County, Delaware. Road No. 261 on Delaware State Highway Department list of 1959.

Mt. Cuba Road Bridge: known as County Bridge #120 near Ashland on Road 261 mentioned above.

Mt. Cuba Bridge: on road southeast of Ashland, no Road #.; possibly bridge #130, which in 1970 was an iron bridge over Barley Mill Road near the Delaware Nature Center.

Wooddale. County Bridge #137 on Road 259A on private property: built ca. 1840, reconstructed in 1969, is still standing; Town lattice truss.

Kiamensi Bridge: County Bridge #156 on Road 331; built ca. 1835-1840, replaced in 1925; Town lattice truss.

Bridge near Brandywine Springs on Hyde Run: built ca. 1840, photograph shows it to be still standing in the 1920s; possibly County Bridge #147 on Road 270; possibly Town lattice truss.

Yorklyn Bridge
County Bridge #112 on Road 257

The Yorklyn Bridge was a one span Town lattice truss structure built over Red Clay Creek on a road joining Mill Creek and Christiana Hundreds. The road and bridge, commissioned by the Levy Court, was opened to traffic on December 4, 1863. The bridge cost approximately $1800. The bridge was unique for two reasons: it was completely enclosed; lattice work on the underside

Yorklyn Bridge. Robert H. Jones Collection.
Courtesy of the Historical Society of Delaware.

Bridges

Ashland Covered Bridge. Photograph by Steven C Gregory Jr., 1995.

and on the sides at each end of the bridge could be seen, and appear to be decorative. According to the New Castle County's Comptroller's records of bridge contracts, the Yorklyn Bridge, County Bridge #112, was slated to be removed and replaced following a September 10, 1929 contract awarded to James H. Hutchison of Newark, Delaware. It was proposed that a steel bridge would replace "the timber covered bridge and will rest upon the present abutments." Curiously, a photograph was taken of the bridge in 1931, and a Delaware State Highway Department photograph shows the Yorklyn Bridge supposedly still standing in the 1950s. The bridge was replaced closer to 1929 than 1950 because a newspaper article of June 1929 concurs that the Yorklyn Bridge's "passing is decreed by the march of progress and the unsentimental automobile." The fact remains, the covered bridge at Yorklyn is no longer there.[43]

The village of Yorklyn was originally known as Auburn, but with the advent of the railroad, the officials of the Landenburg Branch of the Wilmington and Western Railroad chose the name "Yorklyn" for its station there. The Yorklyn bridge should not be confused with Marshall's Bridge located north of Yorklyn just over the Pennsylvania state line. By the 1870s Marshall's paper mills at Yorklyn were producing vulcanized fiber and high quality rag. Brothers Israel W. and T. Ellwood Marshall bought the Auburn cotton factory site and built a new paper mill at the site in 1899 to expand production of these products. Eventually Yorklyn became the home to the National Vulcanized Fibre Company because of the Marshalls' efforts. Various members of the Marshall family erected numerous lucrative mills besides paper along the Red Clay Creek, including iron works and grist mills. The family's name was attached to another village farther south on the creek, Marshallton, so Marshall's bridge eventually became simply known as the Yorklyn Bridge. If any name were to be attached to this area, the John Garrett family, who settled in the area and established grist and flour mills by the mid-1700s, would have been a good choice. This family was just as industrious and successful as the

Marshall family on a smaller scale. The Garrett paper mill, established in Yorklyn by Horatio Gates Garrett, and the first on Red Clay Creek only survived ten years, 1803 to 1813, before being sold eventually to Thomas Lea, of the Lea family merchants at the Brandywine Mills. His nephew, Jacob Pusey, converted it to a successful cotton factory. It was this site the Marshall's purchased at the end of the century. The Garrett family is known for founding the Garrett Snuff Mills in 1782 by John Garrett Jr., an industry unrivaled in the area and is among the longest lived on Red Clay Creek; it lasted well into the 20th century. Garrett's Snuff Mills was the production end of the works; the products were sold through a warehouse in Philadelphia. [44]

Yorklyn was a burgeoning industrial site during the 19th century. The fact that the bridge was not built until millers demanded construction of a new road to connect it to the Newport-Gap Turnpike illustrates that businesses at this local were booming. The picturesque covered bridge at Yorklyn yielded to the strains of this commercial activity, and is just a memory. The news reporter quoted earlier, wrote that mention of the Yorklyn Bridge brought back memories of boyhood days there, particularly "hangovers," where one would hang from the bridge planks. The bridge served as shade and shelter, as well as a community gathering place in the rural village of Yorklyn.[45]

Ashland Bridge
County Bridge #118 on Road 258

The Ashland Bridge on Brackenville Road just off Route 82 is only one of two covered timber bridges still standing in Delaware. It is a Delaware treasure, and because of its history and age the Ashland covered bridge was placed on the National Register of Historic Places in 1973. After Smith's Bridge was burned down by vandals, public action moved to save Ashland and its twin, Wooddale. Ashland is a one span Town lattice truss covered bridge being used currently for the purpose it was originally built, to carry traffic across the Red Clay Creek. It was built, according to Edward F. Heite of the State's Division of Historical and Cultural Affairs in 1974, most likely between 1850 and 1865, although it could have been built as early as 1840. The builder is unknown. The bridge measures 52 feet long with a clear span of 40 feet, and is 14 feet 6 inches wide. The Town lattice work is evident because the diagonal web members of uniform size are crossed at 45 to 60 degrees to form a garden lattice effect. The chords, composed of two or more parallel planks, were spaced allowing the diagonal members to pass between them, creating diamond shaped panels. The overlapping planks were secured originally by hardwood treenails or trunnels. The foundations of the bridge were comprised of rubble masonry; the wingwalls built of uncoursed rubble and topped with sloped capstones. "The portals are ornamented by pilasters with flared capitals reminiscent of classical columns."[46]

Ashland Bridge was on the State Highway Department's list of one "of the worst [structurally] in the state, currently needing attention" in 1974. During 1982 the bridge underwent a major restructuring when rolled steel I-beams were installed under the deck. It reopened late December 1983. Area residents were distressed to see the century old frame hoisted aloft by crane and set alongside the roadway, but rejoiced to see it back in place "atop a lofty steel support system and new road-bed timbers." The three ton weight limit was increased, and the frame painted barn red with white portals. The reinforcement also straightened the bridge at the southern portal which had become noticeably bowed, a common problem with fully timber construction.[47]

This bridge near Ashland was one of five in the area: one was nearby bridge 118; one was east of Red Clay Creek; one west and one southeast on Mt. Cuba Road. Why the concentration of bridges at this location? The paper and snuff milling industries in Yorklyn certainly brought a lot of traffic through the area during the 19th century. The Landenburg Branch of the Wilmington and Western also passed through Ashland, transporting goods and

Bridges

people. Ashland's development into a prosperous mill site had its beginnings in the late 17th century as a result of William Penn deeding the land south of Chester County, including Ashland to his daughter, Letitia Penn, in 1701. In 1702 John Gregg, the son of William Gregg who purchased a tract of land to the northeast at Rockland Manor which became known as Strand Millas, purchased 200 acres of Penn's daughter's tract. The first grist mill to be built on Red Clay Creek was by John Gregg in 1715. Gregg deeded the property to his son William in 1730, who in turn divided the property among his sons, Harmon, William, Joshua and Jacob in 1746, all of whom sold their shares of the estate. John C. and R. Phillips operated the mills between 1790 and 1856 with an annual production of 22,000 barrels of flour and 2800 barrels of cornmeal in 1832. In 1862 Jehu D. and Amos Sharpless bought the property, then being referred to as Ashland Mills. The Sharplesses enlarged their mills and replaced the stones with rollers to make flour. Concurrently, the Ashland Bridge was built,

Covered bridge which once stood on the Ashland Farm of Henry B. du Pont. Richard Sanders Allen, <u>Covered Bridges of the Middle Atlantic States</u> (Brattleboro, VT: The Stephen Greene press, 1959).

as was the Wilmington and Western Railroad by prominent Wilmingtonian Joshua Heald. The Baltimore & Ohio Railroad bought the Wilmington & Western after the B&O constructed a railroad line through the state in 1886.[48]

Ashland industry mirrored what was happening elsewhere on Red Clay Creek in the 20th century. Established businesses remained strong for the most part, but the advent of the automobile and population growth demanded better transportation networks. Some of Red Clay Creek's covered bridges were on highly travelled routes, such as Bridge #118; others fell into disuse, and either collapsed or were washed away by floods. The Ashland Bridge has become a landmark of the craft of bridge building and of the charm such a covered bridge has held and holds for its community.

Bridge East of Ashland
County Bridge #119 on Road 261

This is the first of four bridges on Road 261, also known as Mt. Cuba Road or Route 82. Bridge #119 was built between 1850 and 1865; the builder is unknown. It is of the Town lattice truss type, and stood on the estate of Henry Belin du Pont, his Ashland Farm, which he purchased in 1935. Records show that at one time it had been painted "an eye-knocking red and yellow." Around 1964 Henry B. du Pont and his wife Margaret burned down this bridge because it had become so dilapidated, and they feared it would collapse or be torched by arsonists. This bridge, called the "DuPont Bridge," was bypassed when "road builders straightened a curve in Delaware 82." For over forty years the DuPont Bridge remained unused and unmaintained. It had become so weary, and passersby tore boards off of it. Before anyone was hurt or the bridge fell down the du Ponts had it destroyed. When Henry B. du Pont moved on the property he had a stone bridge built at the entrance to his estate. Because of this bridge and the covered bridge, a friend suggested to Mr. du Pont to name his property "Bridgemoor" or "Stone Bridge Farm." No response was found; the property was known as his Ashland Farm.[49]

Bridges

Mt. Cuba Road Bridge, 1930. Frank Zebley scrapbook. Courtesy of the Historical Society of Delaware.

Mt. Cuba Road concrete bridge, 1995. Photograph by Steven C. Gregory, Jr.

Of the other two bridges one is referred to on a covered bridge list sent to State Archivist Leon deValinger by David K. Witheford of the Delaware State Highway Department in 1959. It reads "Red Clay Creek Bridge at Ashland, New Castle County, Delaware, Road 261." The other bridge is mentioned in Henry B. du Pont correspondence as being washed away in a flood of the creek in July 1938. No other documentation exists of these two bridges, not even photographs. Sorting out the number of covered bridges in this area has been a bit of a puzzle. If indeed these two bridges did exist, they were probably timber Town lattice truss types, since the others around them seem to follow this design.[50]

Mt. Cuba Road Bridge
over Red Clay Creek at Ashland
County Bridge #120 on Road 261

This bridge is the fourth of four bridges on Road 261. It was a timber Town lattice truss structure nearly identical to the extant Ashland bridge. It spanned 59 feet 8 inches and rested on rubble masonry abutments. The bridge was most likely constructed between 1850 and 1865, and was replaced in 1922 by a concrete arch bridge. The covered bridge was not dismantled during the construction of the concrete bridge. Roadway was prepared for traffic between the covered bridge and a piece of land on the Hillside Mills road running between Mt. Cuba and Ashland Stations on the Landenberg Branch of the Baltimore & Ohio Railroad. The covered bridge entrances were boarded up because it was considered dangerous for motor traffic. This did not stop people from using it as a temporary garage and some traffic did get through. In 1934 the Mt. Cuba bridge was torn down.[51]

The geographic proximity of Mt. Cuba to Ashland assured that industry would thrive there, especially with the Red Clay Creek to power any mills established. Transportation routes in the locale were also convenient to deliver raw materials and finished goods. By 1816 Joshua Lobb established a gristmill and sawmill on

Bridges

Woodale Bridge by Charles Colombo. The Cedar Tree Press Collection.

property southeast of Ashland Mills he purchased from Samuel Woodwards in 1814. Twenty years later Lobb sold the land to Miller Speakman. The Speakman family continued to operate the mills until 1873, when they sold it to James Green and James Wilson. Green sold out eventually, and Wilson became sole owner. On a tributary of Red Clay Creek, Burris Run, a short distance from the Mt. Cuba mills, was located a saw mill built by Samuel Graves around 1793. The operation did change hands several times, but was still running in 1894 by Hayes Graves. In the 20th century on the estate of Lammot duPont Copeland in Mt. Cuba was founded the Mt. Cuba Astronomical Observatory, honoring Francis Gurney du Pont, who loved astronomy. The University of Delaware has held courses here in astronomy and astrophysics.[52]

Mt. Cuba Bridge listed as being on the road southeast of Ashland on the 1959 State Highway Dept. bridge list has no road listed for it. It is possibly County Bridge #130, which was an iron bridge in 1970 near the Delaware Nature Center.

Wooddale Covered Bridge
County Bridge #137 on Road 263A
Mt. Cuba

The Wooddale covered timber bridge is called the fraternal twin of the Ashland covered bridge, just two miles away. Both are still standing and on the National Register of Historic Places, although there were no markers in place designating them as such. The bridge now sits on private property, but for over one hundred years it was under the care of the State. Wooddale bridge was a one span covered bridge spanning 53 feet 6 inches long with a 13 foot deck. As drawings of a 1939 Delaware Department of Transportation documentation of Wooddale indicate the bridge's original frame was the Town lattice truss, connected with hardwood dowels, or trunnels built somewhere around 1850, the same as Ashland. "The truss is constructed of 2" x 8" diagonals, and floor beams of 6"x12" timbers. The abutments were of semi-coursed rubble with a smooth finish on the exposed face." The wing walls are flared and are topped with granite capstone. A weight limit of "liveload three tons" is noted. The Wooddale bridge was rehabilitated in 1981 by the property owner. Wood members were added to the floor bracing system, plus a steel I-beam subframe had been added by 1969 to relieve the load stress on the wooden frame. The bridge now "functions as a steel stringer span" with the load limit at five tons.[53]

The owners of the property on which the Wooddale bridge sits have always and continue to be protective of the bridge. It survived the July 1938 flood of the creek, and a torching in March 1975. Because traffic over the bridge is limited, the bridge is spared weight stress. One owner remarked that the bridge was "not quite straight," and that it had possibly been "canted" or "askew" since the mid-nineteenth century. The placement of the I-beams under the bridge by 1969 narrowed the throat of the bridge by twelve feet due to the masonry abutments constructed to hold the I-beams in place. The I-beams in turn lowered the bottom of the bridge twelve inches. This narrowing constricted the flow of water through the throat of the bridge during flooding.[54]

The Wooddale bridge was once a public, not a private, structure as noted above. The bridge has one end in Christiana Hundred and the other in Mill Creek Hundred, and leads off a road connecting Barley Mill Road with Lancaster Pike. Originally a road to the west ran across the bridge and along the edge of the private property until the construction of the Wilmington and Western spur was built across the road, mandating laying of a new road along the east bank of the Red Clay Creek. A siding jutted off to the southwest of the trace, so the bridge allowed access to this siding and to the railroad station, which was once there.[55]

The Wooddale property was once the site of heavy industry, and iron works and paper mills. The second slitting mill in the county opened for operation on the Wooddale site in 1814 by Edward

Gilpin, of the Wilmington papermaking Gilpins, along with a rolling mill. (The first slitting mill was operated by John Gregg at Hagley in 1779.) The slitting mill rolled bars of cast iron into small sheets, which were then slit into barrel hoops, rods from which nails were made, and all purpose strips for blacksmiths. The introduction of nail cutting machinery made obsolete one of the functions of the slitting mill. The technology of puddling cast iron and refining it by running bars through a rolling mill revolutionized iron manufacturing in the United States. In 1826 James Wood and his son Alan leased the rolling mill, and thus the Delaware Irons Works evolved. The impetus for the Woods leasing on Red Clay Creek was the building of the Chesapeake and Delaware Canal. The Woods' specialties at this time were shovels and spades, and many were needed for the canal project. The Woods moved to Conshohocken, Pennsylvania to establish an iron works there, after the Delaware Iron Works property went through two more owners. Alan Wood bought the Delaware Iron Works in 1844, but it took secondary precedence to the Alan Wood Steel Works in Pennsylvania. John Marshall established a rolling mill on the site of a gristmill south of Wooddale. The Marshallton Iron Works gave the Delaware Iron Works competition, on Red Clay Creek as well as Philadelphia since both the Marshalls and the Woods both established ironworks in Philadelphia. The specialty of the Alan Wood Steel Co. was Russian glazed sheet iron; that of the Marshallton Works was galvanized iron. The Delaware Iron Works was hit hard by the 1873 depression, and even though a railroad siding from the Wilmington and Western Railroad in 1875 improved transportation of goods, activity at the ironworks declined. By 1884 the Marshallton Iron Works a few miles south reached a production capacity of four times more than the Woods' operation. The Woods sold the mill in 1891 to Robert Marshall, who converted it to the Wooddale Paper Mill. The Marshall's main paper milling concern was in Yorklyn. Their Wooddale Mill was productive into the 20th century.[56]

Bridges

Kiamensi Bridge over Red Clay Creek, ca 1920s. Delaware State Highway Dept. photograph. Courtesy of the Historical Society of Delaware.

The Wooddale covered bridge saw a lot of traffic cross over its planks, and the area most likely did not appear as picturesque as it does now (1995). This bridge connects us with the industrial past of Red Clay Creek, declaring the importance of continuing to preserve the bridge as long as possible.

Kiamensi Bridge
County Bridge #156 on Road 331

The Kiamensi Bridge, also called the Kiamensi Road Bridge, was a Town lattice truss covered bridge replaced by another bridge type in 1925. A State Highway Department photograph shows the bridge to be a longer bridge than Wooddale or Ashland, and that the lattice frame is completely enclosed with no windows or vents. There is no record of when the bridge was built or by whom, but the period between 1850 and 1865 is when at least four of the covered bridges over the creek were supposedly built, so it would be logical for the Kiamensi Bridge to have been constructed then as well. Length and width of this bridge were not located.[57]

The Kiamensi Woolen Factory operated at the site of cotton mills called the Endeavor Mills owned by Thomas Lea from 1816 to 1826 near Stanton. The name Kiamensi originated in local folklore. Kiamensi was an Indian maiden, who was jilted by her lover, and committed suicide by jumping off high rock into a deep pool in the Red Clay Creek. The woolen factory was one of three in the county to survive any length of time, and it was the first one on Red Clay Creek to incorporate, which was in 1864. The other two were the Auburn cotton mill in Yorklyn, 1813-1880, and the Hyde Run textile mill run by Henry Clark, 1816-1893, both on Red Clay Creek. William Dean, who operated a woolen mill in Newark by 1867, along with five others purchased and resold the Kiamensi factory in 1864 to the newly incorporated Kiamensi Woolen Company. By 1870 the Kiamensi Woolen Factory was the largest manufacturing firm in Mill Creek Hundred. It survived the 1873 depression without heavy losses, and this was basically

Bridges

Bridge near Brandywine Springs, ca 1920s. Delaware State Highway Dept. photograph. Courtesy of the Historical Society of Delaware.

because of the decline of smaller textile concerns, which allowed the larger ones to increase their production. The Kiamensi Woolen Company was still in operation in 1914.[58]

The location of the Kiamensi Bridge near Stanton saw a lot of traffic because Stanton was a thriving market place being on a neck of fastland between the Red Clay Creek and White Clay Creek about two miles from the Christina. Members of the Tatnall and Lea families of Wilmington operated a grist mill in Stanton until 1885 and the Stanton Woolen Company also operated there. After the construction of the Chesapeake and Delaware Canal in 1825, and the advent of railroads in the area after the Civil War, Stanton became less active, but the Kiamensi Woolen Company continued to prosper along with the Henry Clark woolen mill on Hyde Run, site of the next covered bridge.[59]

Bridge near Brandywine Springs Tributary
County Bridge #147 on Road 270

A Delaware State Highway Department photograph is the only documentation found to show that this bridge actually did exist. The bridge is on a covered bridge list compiled by David K. Witheford of the State Highway Department in 1959. The type of construction, when built and by whom, when dismantled or destroyed, and measurements is information unobtainable at this printing, and perhaps forever. A steel truss bridge built in the 1950s stands on or near the site of this covered bridge. The tributary to Brandywine Springs which the covered bridge most likely crossed is Hyde (Hide) Run. The bridge in the state photograph appears to be very similar to the Kiamensi Bridge in that the frame is completely enclosed, and there appears to be no vents or windows. It is possibly a Town lattice truss built between 1850 and 1865, in keeping with all of the other bridges on Red Clay Creek.[60]

This Brandywine Springs area was the locale of two very

different public concerns, one industrial and the other recreational, both founded within ten years of each other. In 1816 John Robinson was operating both a cotton factory and gristmill on the Hyde Run site. By 1834 Henry Clark had acquired the Robinson site and began the manufacture of woolen cloth, which survived the 1857 and 1873 depressions into the 1890s. In 1870 the Henry Clark woolen factory on Hyde Run was the sixth leading mill on Red Clay Creek. Possible relatives of Henry Clark are James Clark, who purchased the Auburn cotton mill in Yorklyn in 1866, and William Clark, who bought it in 1886 only to sell it in 1890 to Israel W. and T. Elwood Marshall. As you may recall, the Marshalls converted the Auburn mill into a paper mill.[61]

The recreational concern on the Springs was operated by the Brandywine Chalybeate Springs Company, a Wilmington partnership who purchased the site in July 1827. The springs, or chalybeate waters, were supposedly health reviving waters containing a strong solution of sulphur and iron. The Brandywine Springs Company envisioned the site as a spa, complete with hotel and picnic grounds, and indeed it became so. It was purchased by Franklin Fell in 1873, whose father established the Fell Spice Mill in 1828. Brandywine Springs became an amusement park in the 20th century complete with a merry-go-round and roller skating rink. In a description of the park in the 1920s a reference is made to "a rustic bridge crossing Hyde Run" approaching the rink, which could possibly be the covered bridge in question here. A stone bridge accommodated spa goers in the first half of the 19th century. Richard Crook, who had managed the Springs in the late 19th century, invested in the Kiamensi Springs Company in 1906. It was located on an acre of land where a fresh water spring had been found on a hill sloping down to the creek just opposite Brandywine Springs. This would be near where Henry Clark had his mill.[62]

The exact location of the covered bridge in relation to the community it served is perhaps something lost to history. The fact remains that it crossed Hyde Run, and a woolen factory and a spa both functioned on this tributary. Since a covered bridge was a gathering place as well as a shelter, its service to the Brandywine Springs community is unquestionable.

Settlement on Red Clay Creek began in 1715 when John Gregg built his grist mill at Ashland. A century later there would be numerous mills constructed along the Creek: seven grist mills, six saw mills, two cotton mills and one snuff mill. Eventually textile manufacture and ironworks would eclipse the other industries. The Newport-Gap Turnpike and the Wilmington-Lancaster Turnpike were constructed to encourage increasing commerce along Red Clay Creek. Some time between 1840 and 1865 ten covered bridges of the Town lattice truss type were built to accommodate the burgeoning businesses. In 1872 the Wilmington and Western Railroad came along the Creek further enabling the transportation of goods. Of these bridges, two of them remain standing as results of intensive preservation. Ashland and Wooddale covered bridges are the only two remaining covered bridges in the County and State, and they are indeed very much a part of our industrial heritage.[63]

Bridges

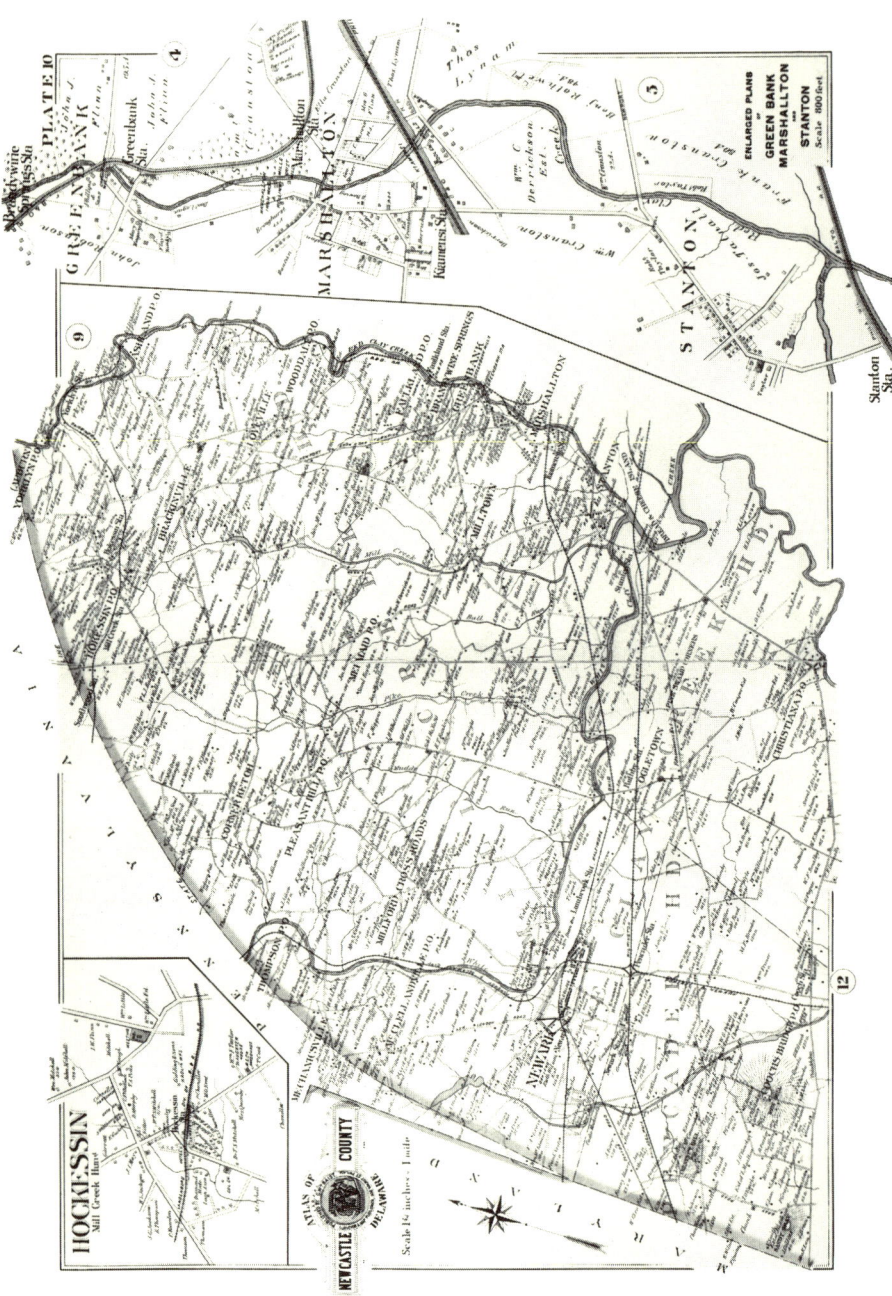

Mill Creek and White Clay Creek Hundreds. G. William Baist, *Atlas of New Castle County* (Philadelphia, PA: G. William Baist, 1893).

Marjorie G. McNinch

White Clay Creek

The White Clay Creek, named for the white clay deposits on its banks, converges with Red Clay Creek and the Christina River near Stanton enclosing land referred to as Bread and Cheese Island. White Clay Creek runs through both Mill Creek and White Clay Creek Hundreds, both areas originally developed from an agrarian economy, but with Mill Creek Hundred getting its name from all the mills dotting the Mill Creek, becoming more industrialized. The town of Newport sat on the important water route between Wilmington and Christiana, destined to become a shipping terminus for Mill Creek and White Clay Creek Hundreds, as well as for southern Chester County, Pennsylvania in the early 1800s. Stanton became a thriving trading center, situated just northwest of the Newport shipping port where the Red Clay Creek and the White Clay Creek converged. With the opening of the Chesapeake and Delaware Canal in 1825, these two towns lost their key positions to this direct water route to Wilmington and Philadelphia.[64]

Stanton, the oldest and once the largest town in Mill Creek Hundred, formerly known as Cuckoldstown, was a thriving trade center in the late 18th and 19th centuries with bridges over both the Red Clay and White Clay Creeks. Grist mill operator William Marshall, whose family became prominent millers and businessmen along Red Clay Creek, resided on White Clay Creek near the bridge crossing. The first mill at Stanton was a saw mill built in 1679. Quakers Stephen Stapler and Samuel Smith erected a grist mill in Stanton in 1800, which was converted into a woolen mill by William Dean, and in 1866 was bought by Wilmington flour merchants Joseph Tatnall and Thomas Lea. A road was constructed in 1768 leading from Stanton to Newark, which had been founded only ten years earlier. Newport, southeast of Stanton,

Bridges

Sketch of wagon being drawn through a covered bridge. Robert Griffith Sketchbook. Courtesy of the Hagley Museum and Library.

offered the nearest shipping route for farmers and millers on Red Clay Creek, White Clay Creek and Mill Creek, who hauled their goods overland by conestoga wagon. By 1825 Newport reached its peak as a shipping terminus. Twenty or thirty wagons at a time carrying wheat, corn, flour and lumber would arrive at the port. The wagon horses would be belled to warn traffic of their coming, and the wagoner would drive from the left side of his team, passing traffic to the right. Our present day custom of driving on the right comes from these wagoners.[65]

The construction of the Chesapeake and Delaware canal in 1825 gave Pennsylvania and Maryland farmers an all water route to Philadelphia, eliminating the need for them to haul goods overland. Business along the Christina River at Newport and Stanton went into a slump. The Pennsylvania Railroad built a line through the area, called the Philadelphia, Wilmington and Baltimore line, before the Civil War. This additional mode of transportation further eroded the towns' prominence as business centers. The needs of area farmers and millers for shipping purposes could not sustain them, so the commercial growth of Newport and Stanton died as a result.[66]

Milling, however, flourished on both Red Clay Creek, discussed earlier, and on White Clay Creek. As along Red Clay Creek, grist and saw mills were the first types of mills along White Clay Creek, saw mills such as those of Benjamin Chambers in 1798 and Thomas and Joseph Rankin, and grist mills such as those of William Marshall, Stephen Stapler and Samuel Smith. John England, who purchased land on White Clay Creek in 1726, also erected a grist mill at the mouth of Muddy Run, which enters White Clay Creek. Other types of mills began to develop. Thomas Meeteer established a paper mill and saw mill west of Newark on White Clay Creek in 1797. In 1804 the Samuel Meeteer and Company paper mill operated under John Armstrong (a Meeteer son-in-law), and son, Samuel Meeteer. The company was known also as the Millford Paper Mills. This establishment was taken over in 1848 by the Curtis family, who still operate the business today

Bridges

White Clay Creek Bridge near Ruthby ca. 1965. Delaware State Highway Dept. photograph. Courtesy of the Historical Society of Delaware.

White Clay Bridge near Harmony before 1925. Delaware State Highway Dept. photograph. Courtesy of the Historical Society of Delaware.

(1995). On land originally purchased by John Guest in 1702, Joseph Dean of Newark established a woolen manufacture in 1848, which became a mainstay in Newark until it burned down in 1886. The Roseville cotton factory near the mouth of Pike Creek where it meets the White Clay Creek operated until it burned in 1868. Iron forges also grew up in both White Clay Creek and Mill Creek Hundred. Iron Hill in Newark provided a vast deposit of iron ore. William Keith, Governor of Pennsylvania purchased the tract in 1722. John England had an iron forge on Mill Creek where it meets the White Clay Creek. Two fulling mills along White Clay Creek were in operation by 1804: William Stapler in Stanton and Joshua Johnson. Tan yards also developed as an industry in this agricultural community. Robert Crawford had one on Muddy Run.[67]

The two Hundreds which shared White Clay Creek needed numerous bridges to carry traffic and goods between them. Goods and raw materials were also hauled overland to the Chesapeake and Delaware canal once it opened in 1825 along the southern edge of New Castle County. The covered bridges over White Clay Creek for which there is documentation number five. No structural documentation about these bridges was located, however, but some Delaware State Highway Department photographs and lists saved these covered bridges from oblivion. The bridges are as follows.

White Clay Creek Bridge near Harmony: neither builder nor date is known; the bridge still standing in 1925; appears to be a single span; bridge type unknown; County bridge #240 on Road 355.

White Clay Creek Bridge near Ruthby Station: neither builder nor date is known; bridge still standing in 1925; bridge type unknown; County bridge #236 on Road 352.

Paper Mill Bridge (Curtis Paper Mill Bridge): built 1861; replaced in 1948-49 by a concrete bridge; appears to be a single span bridge; builder and bridge type unknown; County bridge #231 on State Road 72 and County Road 13.

White Clay Creek Bridge at Thompson's Station: neither builder nor date is known; still standing in the 1920s; bridge # unknown, on County Road 329.

Yeatman's Bridge: built in 1874; builder unknown; bridge burned down in 1960; bridge type unknown; bridge # and County road unknown.[68]

White Clay Creek Bridge near Harmony
County Bridge #240 on Road 355

This covered wooden bridge was located on the south side of White Clay Creek in White Clay Creek Hundred near where the the mouth of Pike Creek enters White Clay Creek. The Harmony Mills, a grist milling operation near the White Clay Creek at this intersection, appears in 1868 and 1893 maps of the area. The bridge, from the photograph, appears to be a single span bridge with square portals. The bridge is enclosed on both sides with no vents nor windows cut into them. The abutments appear to be rubble masonry, and the wingwalls also. The Pennsylvania Railroad line ran south of the Harmony Mills. The mills and the bridge most likely were located on Harmony Road near Route 2. A bridge was built at this location in 1832, and cost $1700. This could have been the Harmony covered bridge, but no designation as to type of bridge built makes it uncertain.[69]

White Clay Creek Bridge near Ruthby Station
County Bridge #236 on Road 352

This covered wooden bridge was located in White Clay Creek Hundred on the south side of White Clay Creek in between where Pike Creek and Muddy Run flow into White Clay Creek. It was located on Red Mill Road near Route 2. The bridge may be a

Postcard showing the covered bridge over the White Clay Creek on Paper Mill Road coming from Newark, 1909. The bridge was replaced in the late 1940s. (Courtesy of the Historical Society of Delaware.)

single span. In the photograph it appears enclosed on both sides with no windows or vents cut in them. The abutments appear to be rubble masonry. On the Baist's 1893 map of Mill Creek and White Clay Creek Hundred, the designation Red Mill Bridge is marked at this spot over the White Clay Creek. Theordore Ruth's name appears close by, and his name was adopted for the Ruthby Station of the Pennsylvania Railroad line which ran through Ogletown. John Ogle, for whom Ogletown is named, was one of the first settlers in the area. In 1667 he settled in New Castle, but purchased large tracts of land in different parts of the county. The land he bought in Mill Creek Hundred along White Clay Creek in 1683 was a 430 acre tract. His son Thomas added to this tract, which extended into Newark, in 1739. This covered bridge served an area largely settled by 1800. Farmers and millers would use this bridge to haul their goods to the nearest turnpike by 1815, using either the New Castle-Frenchtown turnpike, which went southwest, the Newport-Gap turnpike, which went north or the

Wilmington-Lancaster turnpike, which went into Wilmington. When the railroad line went through after the Civil War, the railroad transportation route was a faster route by which to ship goods. The covered bridge, here again, was a necessary link. The fact that it survived into the 1920s speaks to this need.[7]

Paper Mill Bridge
(also known as the Curtis Paper Mill Bridge)
County Bridge #231 on State Road 72 & County Road 13

The Paper Mill Bridge on Curtis Mill Road leading to the Curtis Paper milling complex was a timber covered bridge built in 1861 over White Clay Creek. The bridge was ultimately replaced in 1949, although plans existed for its replacement as early as 1932. In a May 4th letter between the County Engineer and Robert C. Levis of Curtis & Brother Company was a discussion of the size of the bridge to replace the wooden covered bridge. It was anticipated to be a three span bridge. A news article running in January 1940 said the seventy-nine year old bridge was on the retired list. The covered bridge was ultimately replaced in 1949 by a concrete, rigid frame bridge 170 feet long and 26 feet wide. This bridge provided for two four foot wide sidewalks. Specifications and bridge type of the covered bridge has not been determined. From the photographs it appears to be a single span bridge, as most covered bridges dating from the 19th century tended to be constructed. The wingwalls appear quite flared, and the bridge entrance square. The bridge frame was enclosed on all sides with no windows or vents cut into them. The demands of traffic and commerce dictated its demise. Paper Mill Road was widened because of the growth of the City of Newark and the college community there.[71]

The Curtis Paper Company, now owned by the James River Corporation, is located on a paper milling site established in 1789 by Thomas Meeteer. Thomas Meeteer, his sons Samuel, William and George, and son-in-law John Armstrong basically operated the paper mill in Newark from 1789 until Samuel's death in 1838. The Curtis Brothers, George B. and Solomon M., of Philadelphia were familiar with papermaking and were looking for a site when they found the Meeteer property in 1848 in disrepair. By 1850 the Curtis concern was well into operation. A new steam engine was installed, and chemically treated wood pulp as well as rags were used to produce paper. During and after the Civil War, Curtis Paper was run with government contracts almost exclusively. Its principle products since then have been envelopes, cards and fine color paper. Possibly because of the demands on the Curtis Paper mill by the government during the Civil War, the company was prompted to erect a sturdy covered bridge to its site. In 1816 a bridge was built over Tyson's Ford near Meeteer's Mills, but the covered bridge was built to serve the paper company's comings and goings. The old paper mill was razed in 1887 and a new one erected in its place. The Curtis Paper Company underwent several reorganizations, the first one being in 1911. The James River Corporation took ownership of Curtis Paper in January 1977. The Paper Mill covered bridge was one of six remaining covered bridge standing when it was dismantled in 1949.[72]

White Clay Creek Bridge near Thompson's Station
On Road 329

Thompson's Station was a depot on the Pennsylvania Railroad's line running through White Clay Creek Hundred. Not much is known of the covered bridge crossing White Clay Creek at this point. The bridge's dates, its builder, and design type are unknown. A Delaware State Highway Department photograph is the main surviving record of this bridge. Like the other bridges crossing White Clay Creek, Thompson's Station bridge appears enclosed on both sides without windows or vents cut into them. It is most likely a single span. The photograph of the bridge dates in the 1920s. The bridge was located on the road bordering the Walter S. Carpenter State Park.[73]

Bridges

White Clay Creek Bridge near Thompson's Station, ca 1920s. Delaware State Highway Dept. photograph. Courtesy of the Historical Society of Delaware.

This locale was the site of a saw mill and grist mill. Benjamin Chambers built a saw mill on White Clay Creek on land northwest in Mill Creek Hundred by 1798. By 1843 Daniel Thompson bought the land and mill, and built a grist mill on it. The mills stopped operation in 1881, and the property was owned then by Joel Thompson of Newark. Because this site was near where the White Clay Creek flowed from Pennsylvania, Thompson's Station was along the railroad route out of Pennsylvania going south. The milling community had Corner Ketch, Milford Crossroads and Newark as neighbors, so was an important link in the communication chain. The bridge served these neighboring villages by providing access across the creek to the railroad line.[74]

Yeatman's Bridge
No County Bridge or Road Numbers

The Yeatman timber covered bridge was one of five remaining covered bridges standing in New Castle County when it was burned down by vandals in November 1, 1960. The bridge was built in 1874 by Theordore Bird. It was located crossing White Clay Creek near the Delaware line northwest of Thompson's Station. No Delaware State Highway Department photograph documents its existence, however. No bridge number, road number, or bridge specifications have been located. The only references to it are found in a 1991 Delaware Bridge Survey, and two newspaper articles. The November 21, 1960 article describing the "newest bridge in the nation", the Westminster Covered Bridge, mentions the deep sadness felt when Yeatman's bridge, still in service, was torched. The 1974 news report simply refers to Yeatman's Bridge as the last covered bridge to span White Clay Creek, and bemoans its loss to fire.[75]

The bridge most likely took its name from Thomas Yeatman, an early settler who purchased 150 acres from the William Penn Tract "Stanling Manor" located on the circle (boundary) on

September 15, 1723. In a 1936 history of Newark, Yeatman's (assuming it to be a mill or village) is mentioned as being west of Corner Ketch, that a road existed between the Mill Creek Meeting House and Yeatman's. Another reference in the same work puts Yeatman's in London Britain Township on White Clay Creek in Pennsylvania. The Yeatman's referred to here was a mill. B. Yeatman appears on Beers' 1868 map of Mill Creek Hundred on the Delaware/Pennsylvania border west of Hockessin.[76]

The covered bridges over White Clay Creek have not been as well documented as those on Brandywine Creek and Red Clay Creek, but their service to their communities was just as important. The bridges all appear in the photographs to be of similar construction, just as all the bridges along Red Clay Creek were of the Town Lattice truss type, and those spanning the Brandywine Creek were of the Burr arch type. Bridge builders generally were expert carpenters and belonged to a unique group in their trade of bridge building. Little competition existed among bridge builders because of their bridge building knowledge, and each one controlled a certain territory. This certainly holds true along the New Castle County creeks. The builder constructing bridges along each of the creeks duplicated the favored design the length of the creek. Similarly, the bridges along each of the creeks seem to be built all in the same decade. Most of those along Brandywine Creek were erected in the 1830s; those along Red Clay Creek conjectured to be between 1850-65. Since there are no construction dates for three out of the five on White Clay Creek, it can logically be assumed that the three were built between the time the Paper Mill covered bridge saw light, 1861 and Yeatman's in 1874. Along all three creeks, grist mills and saw mills were the first establishments along their banks. The Brandywine had the most pronounced decline to tidewater, so its power drew ambitious and talented millers early to its shores, and made Wilmington the favored port until it was eclipsed by Philadelphia.

The water power of both Red Clay Creek and White Clay Creek also powered industries of early settlers on their banks as well, but these areas tended to be less populated and more agrarian. The covered bridge heritage of New Castle County is one befitting the First State, and some of the spans rivaled those in the Northeast and Mid-Atlantic.[77]

Bridges

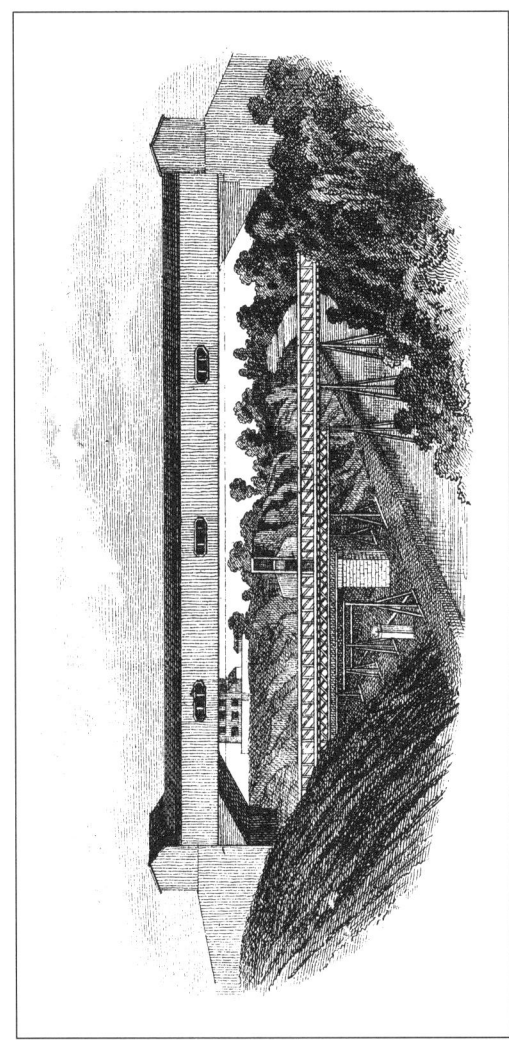

Vignette of the covered bridge over the Chesapeake & Delaware canal decorating a 1908 Chesapeake & Delaware Canal company stock certificate. Loaned by Terry A. Bryan, D.M.D., Dover, DE.

Marjorie G. McNinch

One of a Kind Covered Bridges

Summit or Buck Bridge

New Castle County sported one of the most unique covered bridges in the United States during the 19th century: the Summit, or 'Buck', Bridge over the Chesapeake and Delaware Canal. The Swedes who settled in Wilmington conceived the idea of connecting the Chesapeake Bay and Delaware River as early as 1654. Land surveys were conducted in 1786 to determine the feasibility of such a venture. The Maryland General Assembly chartered the Chesapeake and Delaware Canal Company in December 1799. By 1803 the organization of the Chesapeake and Delaware Canal Company was completed with representatives from Delaware, Maryland and Pennsylvania being chosen; the company was incorporated that same year. They chose a route via the Elk River, and work on the canal was begun in 1804. At the peak of construction early in 1826 the C&D Canal Company employed 2600 men. The town of Delaware City grew as a result of being the terminus of this canal. The bridge over the canal, at Buck Tavern, was finished in October 1826, but the canal was not formally opened to water traffic until October 17, 1829. The canal originally consisted of three locks, each 100 feet long and 22 feet wide with a depth of ten feet. It cost $2,201,864, the tolls of which helped pay for it. The canal was enlarged in 1855 and again in 1935, and is one of the only canals built in the 19th century to still be in operation.[78]

The bridge over the canal was described by civil engineer Henry Tudor in 1833 as "a bridge of singular appearance and ingenious construction." The bridge was 225 feet long and nearly 90 feet above the surface of the water. It straddled the "Deep Cut" in the canal created by men digging and blasting through solid rock nearly a mile long. This covered bridge had ornate windows cut in the sides, and boxed portals. Because of its position high above the

Bridges

Covered bridge built in 1961 at the entrance to the development of Covered Bridge Farms in Newark, DE. Photograph by Steven C. Gregory Jr., 1995.

canal Summit Bridge, also known as Buck Bridge because of the Buck Tavern, became a beacon as well as a landmark to users of the waterway. The bridge did not require a bridge tender, but in 1872 it was replaced by a swing drawbridge at a lower level, which did require a bridge tender.[79]

The job as bridge tender was held by the same family for fifty years. The first bridge tender was John Fletcher Kane, who tended from 1872 to 1905; his son Wallace served as tender from 1905 to 1923. The bridge tender and family were provided with a house. The village of Summit Bridge in St. George's Hundred of New Castle County, being on the south side of the canal, took its name from the covered bridge. In 1888 it consisted of a church, a post-office, the Delaware Wagon Works, two blacksmith shops, a hotel, three stores, and fifteen residences. The Buck Tavern in Summit Bridge, for which the covered bridge was also named, was alleged to have been visited by George Washington three times. A building believed to have been the Buck Tavern was razed in 1963. Mrs. Wallace Kane said she had lived in the village all her life and that the building razed was only a residence, never a tavern. The year 1821 was found carved in one of the planks, bringing an element of credence to Mrs. Kane's words. Both because of visits by the first President of the United States, and because of the Summit covered bridge, this village has reason to be proud.[80]

Westminster Bridge.
Off Route 41, Hercules Road.

The November 21, 1960 newspaper headlines read "Covered Bridge is Nation's Newest" and "Newest Covered Bridge has some enduring features." The bridge to which these articles referred was located just off the Lancaster Pike at Hercules Road southwest of Wooddale, and spanned Hyde Run, a tributary of Red Clay Creek. Hyde Run was crossed by a covered bridge in the 19th century near the woolen mills of Henry Clark and Brandywine Springs, so precedent had been set. The builder and designer of 'this newest covered bridge in the nation' was local contractor Emilio Capaldi, who erected the bridge for the new community of Westminster. The bridge is a single span stringer covered bridge, 32 feet long. The bridge made it into the covered bridge publication of the National Society for the Preservation of Covered Bridges, Inc. in 1960 and 1972. The bridge is listed as being over Coffee Run. The Society declared the bridge to be the newest in the nation as of November 1960. In 1989 the Society put out a guide to replicated covered bridges, since quite a few modern ones were being built, and Westminster Bridge appears in it. Their main guide is strictly for authentic covered bridges. The bridge is also linked to the Delaware Memorial Bridge. Some steel mat wire left over from the construction of the Delaware Memorial Bridge was used to reinforce the concrete floor of the covered bridge.[81]

The specifications of the Westminster Bridge are as follows. The length is 16 feet between the banks of Hyde Run, 32 feet overall, 22 feet wide and 12 feet high. Windows run along the top of each side just below the roof overhang. The cost was $17,000. The reinforced concrete floor is a 20th century feature for the covered bridge. Mr. Capaldi fulfilled a life-long ambition to build a covered bridge, and was surprised at the national attention he received. The Westminster bridge was an important project for covered bridge lovers because it was a reconstruction of one of a America's beloved historic landmarks.[82]

Covered Bridge Farms Bridge.
Off Route 273, Wedgewood Road

Like Westminster, the covered bridge built over the Christiana River off Route 273 and Wedgewood Road was constructed at the entrance of the community which was developed around it. The developer and bridge designer was Harlan C. Williams, who conceived of the idea for New Castle County's second new covered bridge in order to give a name to his community. Mr. Williams was partly inspired by Mr. Capaldi, and partly by Mr. Williams' father, who played a pivotal role in saving an old Cecil County, Maryland covered bridge. The Covered Bridge Farms bridge crosses the east branch of the Christiana River outside of Newark, and was built in 1961. It is a single span stringer type covered bridge, measuring 60 feet long. The Covered Bridge Farms covered bridge was also listed in the National Society for the Preservation of Covered Bridges world guide in 1972, and also in its 1989 new covered bridge guide. It is also the only covered bridge known to have crossed the Christina River.[83]

Marjorie G. McNinch

Market Street mills and covered bridge illustration on the 1859 sheet music cover for the "Brandywine Polka." Loaned by Terry A. Bryan, D.M.D., Dover, DE.

Marjorie G. McNinch

The Art of the Covered Bridge

The covered bridges of New Castle County one by one succumbed to age, neglect, heavy traffic or arson. The news headline "Another Covered Bridge Doomed by Levy Court," written in 1934 is echoed by one written forty years later, "Neglect, vandals endangering our historic covered bridges," 1974. These news headlines bring attention to the demise of the covered bridge in our state. The formation of the Historic Red Clay Valley occurred two years after the 1974 article for the specific purpose of preserving historic sites in the Red Clay Creek Valley, including the two authentic covered bridges still standing, Ashland and Wooddale. The covered bridge of 19th century New Castle County added a quaint charm to "many a winding stream" such as the Brandywine, Red Clay and White Clay Creeks. Covered bridges were "places to be explored" by young boys, shelters in a storm, private spots for lovers, and community gathering places during the 19th and early 20th centuries. The advent of the automobile doomed the covered bridge, which was not built to handle loads more than 3 to 5 tons. The weight of the automobile was not the only feature about it to doom the covered bridge; it carried more people, more places, more often, so traffic across the covered bridge was heavier and more frequent. Communities around the bridges grew, and the new populace did not hold the sentiment the earlier one did. Timber planks in the flooring would break through, planks would be pulled off the sides, the bridge would begin to lean, and then the county engineer would step in and have it replaced. Replacement was a necessary, but a sometimes sad event, as the newspaper headlines attest. So one by one the county's bridges disappeared.[84]

Fortunately New Castle County's covered bridges were captured on film, but they were also captured on canvas by some of Delaware's renowned artists. Images of covered bridges also graced sheet music, stock certificates, and paper currency. The paintings hold the nostalgia of yesteryear. The well known painting, "Brandywine Mills," by Bass Otis in 1840 shows the Market Street covered span linking the mills. Frank Jefferis' painting "The Brandywine" is a scene of Henry Clay Village near the DuPont Powder Works, and in the center is the Rising Sun covered bridge. John Rubin Smith painted the Rockland covered bridge. Stanley Arthurs' painting of Thompson's Bridge was one of twelve selected for the DuPont Company's 1941 Safety Calendar. Samuel Homsey's "The Covered Bridge" was one of many valued paintings to hang in the offices of the DuPont Building in the 1950s. Celebrated artist Robert Shaw did an etching of the old North Market Street Bridge, now housed at the Historical Society of Delaware. Because Wooddale and Ashland covered bridges are the only two in the county still standing since Smith's Bridge was burned down in 1961, numerous Delaware artists have sketched and painted them. Wooddale has been the subject of paintings by Charles Colombo and Richard C. Layton. Ashland has been illustrated by Paul Scarborough. Wooddale and Ashland were sketched by artist Nancy Sawin, and appear in her publication *Delaware Sketch Book*. As one news reporter wrote, Delaware's covered bridges are surrounded "with a halo of sentiment."[85]

The sentiment for the charming covered bridge spilled over into other artistic forms. The Brandywine Mills Bridge (the North Market Street Bridge) graces the cover of the "Brandywine Polka" sheet music, written by E. Manuel and published by Duffey & Miller in 1859. The caption under the rendering says "drawn from nature." Covered bridges were symbols of commerce and industry in the 19th century, and as such were imprinted on paper currency and stock certificates. The love for covered bridges will remain. "Covered bridges go but memories live."[86]

Memories of the old covered bridge at Rising Sun were committed to paper in another art form, the poem. Edward B. Cheney, DuPont powderman, wrote the following words not long before the covered bridge was replaced in 1928. These words keep the charm of "the plank pin bridges" alive.

Marjorie G. McNinch

THE OLD COVERED BRIDGE

Oh, old covered bridge o'er the Brandywine,
At the foot of Rising Sun Hill,
With holes in your roof, and unpainted sides,
And broken window sill.
With dusty rafters where sparrows chirp,
And uneven planks on the floor,
You're one of the landmarks left on the 'Crick,'
To remind us of days of yore.

You need some attention, dear old bridge,
A coat of nice fresh paint,
A new shingle roof, your sides repaired,
You're enough to distress a Saint.
It won't be long till you'll be torn down,
And a new bridge take your place,
You are one of the few that is left us,
Like the last of a vanishing race.

Edward B. Cheney

Ashland Covered Bridge by Paul Scarborough. The Cedar Tree Press collection.

Marjorie G. McNinch

BIBLIOGRAPHY
Published Works

BOOKS:

Allen, Richard Saunders. *Covered Bridges of the Middle Atlantic States*. Brattleboro, VT: The Stephen Greene Press, 1959.

Allen, Richard Sanders. *Covered Bridges of the Northeast*. Brattleboro, VT: The Stephen Greene Press, 1974.

Automobile Club of Delaware County. *Blue Book*. Vol. 3, 1916.

Baist, G. William. *Atlas of New Castle County*. Philadelphia, PA: G. William Baist, 1893.

Bartlett, John. *Familiar Quotations*. 14th edition, revised and enlarged. Emily Morison Beck, Editor. Boston and Toronto: Little, Brown and Company, 1973.

Beers, Daniel G. *Atlas of the State of Delaware*. Philadelphia, PA: Pomeroy & Beers, 1868.

Bryant, William Cullen, editor. *Picturesque America*. 2 vols. New York: D. Appleton & Company, 1872-1874.

Brydon, Norman F. *Of Time, Fire and the River: The Story of New Jersey's Rivers*. Essex Fells, New Jersey: 1971.

Committee on History and Heritage Of American Civil Engineering. *American Wooden Bridges*. ASCE Historical Publication #4. New York: American Society of Civil Engineers, 1976.

Conrad, Henry C. *History of the State of Delaware from the Earliest Settlements to the Year 1917*. 3 vols. Wilmington, Delaware: 1908.

Cooch, Francis A. *Little Known History of Newark, Delaware and Its Environs*. Newark, DE: The Press of Kells, 1936.

Cooper, Constance J. *The Curtis Paper Company: From Thomas*

Meeteer to the James River Corporation. Wilmington, DE: The Cedar Tress Press, Inc., 1994.

Delaware Department of Transportation, Geographic Information Section. *General Highway Map—New Castle County*. Dover, DE: DELDOT, 1992.

Fitzsimons, Gregory G. *Historical Bridge Survey of New Castle County: Project Report*. Graduate course requirement in the Civil Engineering Department, University of Delaware, under review by Dr. Eugene Chesson. Newark, DE: University of Delaware, 1982.

The Hagley Museum Guide. Wilmington, DE: Eleutherian Mills-Hagley Foundation, 1976.

Jamison, Kathleen M. "The Death and Life of Red Clay Creek." *Delaware Conservationist*, Delaware Department of Natural Resources and Environmental Control, Vol. XXXI, No. 2, 1988.

Kelly, Leslie A. "A Wealth of Bridges." *Country*, April 1985.

Kuhlmann, Charles Byron. *The Development of the Flour-Milling Industry in the United States*. Boston and New York: Houghton Mifflin Co., 1929.

MacElree, Wilmer W. *Along the Western Brandywine*. 2nd ed. West Chester, PA: F. S. Hickman, 1912.

MacElree, Wilmer W. *Down the Eastern and Up the Black Brandywine*. West Chester, PA: F. S. Hickman, 1906.

Monroe, John A. *History of Delaware*. A University of Delaware Bicentennial Book. Newark: University of Delaware Press, and London: Associated University Press, 1979.

Montgomery, Elizabeth. *Reminiscences of Wilmington in Familiar Village Tales*. Philadelphia, PA: T. K. Collins, Jr., 1851.

National Society for the Preservation of Covered Bridges. *World Guide to Covered Bridges*. Revised edition. South Peabody, Mass.: 1972.

Pursell, Caroll W., Jr. "The Delaware Iron Works, A Nineteenth Century Rolling Mill." *Delaware History*, Vol. VIII, No. 3, March 1959. Wilmington, DE: Historical Society of Delaware, 1959.

Pursell, Carroll W., Jr. *Ironworks on Red Clay Creek in the 19th Century: The Wooddale and Marshallton Mills of New Castle County, Delaware*. Wilmington, DE: Historic Red Clay Valley, Inc., 1962.

Pursell, Carroll W., Jr. *That Never Failing Stream: A History of Milling along Red Clay Creek during the nineteenth century*. Master's Thesis. Newark, DE: University of Delaware, June, 1958.

Rasmussen, William C. "The Water Resources of Northern Delaware," Delaware Geological Survey, *Bulletin*, no. 6 (June 1957).

Sawin, Nancy and Carper, Janice. *Delaware Sketch Book: An Historical Experience*. Hockessin, DE: The Holly Press, 1976.

Scharf, J. Thomas. *History of Delaware, 1609-1888*. 2 vols. Philadelphia, PA: L. J. Richards & Co., 1888.

Spero, P. A. C., & Company. *Delaware Historic Bridges Survey and Evaluation*. Delaware Department of Transportation Project 87-07002; Historic Architecture and Engineering Series No. 89. Prepared for the Delaware Department of Transportation. Dover, DE: Delaware Department of Transportation, 1991.

[Taylor, Frank H.] *History of the Alan Wood Iron and Steel Company, 1792-1920*. Philadelphia: 1920.

Welsh, Peter C. "The Brandywine Mills: Chronicle of an Industry, 1762-1816." *Delaware History*, V. 11, 1956. Wilmington, DE: Historical Society of Delaware, 1956.

Weslager, Clinton A. *Brandywine Springs, The Rise and Fall of a Delaware Resort*. Wilmington, DE: Hambleton Company, Inc., 1949.

Weslager, Clinton A. *Delaware's Forgotten River, The Story of the Christina*. Wilmington, DE: Hambleton Company, Inc., 1947.

Weslager, Clinton A. *140 Years Along Old Public Road*. Wilmington, DE: Historic Red Clay Valley, Inc., 1960.

Williams, Henry William. *The First State: An Illustrated History of Delaware*. Produced in cooperation with the Delaware State Chamber of Commerce. Northridge, CA: Windsor Publications, Inc., 1985.

Zacher, Susan M. *The Covered Bridges of Pennsylvania, A Guide*. Harrisburg, PA: Pennsylvania Historical and Museum Commission, 1982.

Zebley, Frank. *Along the Brandywine*. Wilmington, DE: Lithographed by W. N. Cann, Inc. 1940.

NEWSPAPERS:

Hockessin Community News, Hockessin, DE, 1970s.

Journal Every Evening, Wilmington, DE, 1936-1965.

Morning News, Wilmington, DE, 1957-1974.

Niles Weekley Register, Vol. IX (1815).

Pennsylvania Gazette, August 23, 1764.

Philadelphia Inquirer, Philadelphia, PA, 1974.

Sunday Star, Wilmington, DE, 1929-1934.

Unpublished Works

Federal Writers Project, Del. F164F47, Vols. 7, 16, 24 and 35. Special Collections, Morris Library, University of Delaware.

Hancock, Harold C. "The Industrial Worker Along the Brandywine, 1800-1900." 3 vols. Wilmington, DE: Eleutherian Mills-Hagley Foundation, August, 1958. [Hagley Museum and Library, Accession 1645 #35].

Manuscripts and Archives Collections

"The Brandywine Mills," by Samuel C. Rumford. Accession 601, Hagley Museum and Library.

Comptroller. Roads and Bridges contracts and specifications, New Castle County. RG 2450, Delaware State Archives.

Covered bridges, General Reference File 565, Delaware State Archives.

Delaware Department of Transportation Bridge Files, Dover, DE.

Delaware State Highway. Contracts Index, one volume. RG 1540, Delaware State Archives.

Henry Belin du Pont Papers, Accession 1608, Box 20 and 110, Hagley Museum and Library.

William P. Frank Collection, Box 4, Historical Society of Delaware.

Robert H. Jones Collection and photograph album, 85.9, Historical Society of Delaware.

New Castle County Bridge Book, 1913. RG 2460, Delaware State Archives.

Irving Warner Collection, Accession 1518, Box 16. Hagley Museum and Library.

Frank Zebley Photo Albums, Historical Society of Delaware.

Terry A. Bryan, D.M.D., Dover, DE, Private collection of Delawariana.

Marjorie G. McNinch

FOOTNOTES

[1] Federal Writers Project (Del. F164F47), Vol. 7, pp. 33, 225, Special Collections, Morris Library, University of Delaware (hereafter Federal Writers Project, UofD; J. Thomas Scharf, *History of Delaware, 1609-1888*, Vol. 1 (Philadelphia: L. J. Richards & Company, 1888), pp. 37-57; Kathleen M. Jamison, "The Death and Life of Red Clay Creek," in *Delaware Conservationist* (Dover, DE: Department of Natural Resources, 1988), Vol. XXXI No. 2, pp. 5; William Henry Williams, *An Illustrated History of the First State, Delaware* (North Ridge, California: Windsor Publications, Inc.), pp. 18-30, 62, 71-73. See also John A. Munroe, *History of Delaware* (Newark, DE and London: University of Delaware Press and Associated University Press, 1979).

[2] Committee on History and Heritage of American Civil Engineering, *American Wooden Bridges*, American Society of Civil Engineers Historical Publication #4 (New York: American Society of Civil Engineers, 1976), pp. 14, 126.

[3] ASCE, *American Wooden Bridges*, p. 3.

[4] Ibid, pp. 3, 15, 46.

[5] Ibid, p. 3.

[6] P. A. C. Spero and Company, *Delaware Historic Bridges Survey and Evaluation*, Historic Architecture and Engineering Series No. 89 (Dover, DE: Delaware Department of Transportation, 1991), pp. 170-172; C. A. Weslager, *Delaware's Forgotten River: The Story of the Christina* (Wilmington, DE: Hambleton Company, Inc., 1947), pp. 152-153.

[7] Spero, *Historic Bridges*, pp. 172-173; Carroll Wirth Pursell, Jr., *That Never Failing Stream: A History of Milling along Red Clay Creek during the Nineteenth Century*, Master of Arts Thesis requirement (Newark, DE: University of Delaware, June 1958), pp. 48-49.

[8]New Castle County Bridge Book, 1913, Delaware State Archives (hereafter State Archives), Dover, Delaware.

[9]ASCE, *American Wooden Bridges*, pp. 48, 127; Nelson G. Brooks, "Covered Bridges Go But Memories Live," in the *Wilmington Morning News* (hereafter MN), Wilmington, DE, August 24, 1957, Vertical File titled "Delaware Covered Bridges," Wilmington Institute Free Library (hereafter WIFL, Wilmington, DE; "But Why Covered?", no author, Vertical File titled "Covered Bridges," Cecil County Public Library, Elkton, MD.

[10]ASCE. *American Wooden Bridges*. pp. 126-127; Brooks, "Covered Bridges Go," MN 8/24/1957, WIFL; Leslie A. Kelly, "A Wealth of Bridges," in *Country*, April 1985, p. 62.

[11]ASCE, *American Wooden Bridges*, pp. 47, 97; Spero, *Historic Bridges*, pp. 22-23, 170; Gregory Fitzsimons, *Historical Bridge Survey of New Castle County, Delaware: Project Report* (Newark, DE: University of Delaware, 1982), p. 5; Kelly, "Wealth," p. 62.

[12]ASCE, *American Wooden Bridges*, pp. 14, 50, 125, 156.

[13]Ibid, pp. 46-47, 152.

[14]Ibid, pp. 15, 157.

[15]Ibid, pp. 14, 121, 126.

[16]Brooks, "Covered Bridges Go," MN 8/24/1957, WIFL; no author, "Will Soon Be a Fragrant Memory," *Sunday Star* (hereafter SS), Wilmington, DE, August 12, 1934:1, Magazine Section, WIFL.

[17]"Fragrant Memory," SS 8/12/1934:1, WIFL; Federal Writers Project, Vol. 24, pp. 48, UofD; Brooks, "Covered Bridges Go." MN 8/24/1957, WIFL; Henry Hamm, "Camera Clicked—And Wilmington Man Started on a Hobby," *Journal Every Evening* (hereafter JEE), September 20, 1941:2, WIFL.

[18]Kelly, "Wealth," p. 62; ASCE, *American Wooden Bridges*, p. 128; Brooks, "Covered Bridges Go," MN 8/24/1957, WIFL; Betty

Harrington MacDonald, "Historic Landmarks of Delaware and the Eastern Shore," JEE April 2, 1955:14, WIFL.

[19]Elizabeth Montgomery, *Reminiscences of Wilmington, in Familiar Village Tales, Ancient and New* (Philadelphia: T. K. Collins, Jr., 1851), pp. 8, 29; William Cullen Bryant, ed., *Picturesque America*, 2 volumes (New York: D. Appleton & Company, 1872-1874), Vol. 1, pp. 220-22; Wilmer W. MacElree, *Down the Eastern and Up the Black Brandywine* (West Chester, PA: F. S. Hickman, 1906), p. 18.

[20]Fitzsimons, *Historical Bridge Survey*, Survey Reports after p. 9; 1959 Delaware Covered Bridges list by David K. Witheford, Planning Engineer, Delaware State Highway Department, General Reference File 565 titled "Covered Bridges," State Archives; 1959 "Covered Bridges of Delaware" list by Richard Sanders Allen, Gen. Ref. File 565, State Archives.

[21]Letter, Richard Sanders Allen to David K. Witheford, March 25, 1959, Gen. Ref. File 565, State Archives.

[22]Fitzsimons, *Historical Bridge Survey*, pp. 7, 9; Pursell, *That Never Failing Stream*. p. 5; Frank Zebley, *Along the Brandywine*, (Wilmington, DE: Lithographed by W. N. Cann, Inc. 1940), p. 95.

[23]Federal Writers Project, Vol. 24, p. 82, UofD; Zebley, *Along the Brandywine*, p. 95; Scharf, *History of Delaware*, Vol. 2, pp. 907; 1959 Witheford and Allen bridge lists, Gen. Ref. File 565, State Archives; Fitzsimons, *Historical Bridge Survey*, Smith Bridge Survey Report; no author, "100 Year-Old Smith's Bridge Over Brandywine to be Repaired, Preserved," MN December 17, 1954:1, WIFL; MacDonald, "Historic Landmarks," JEE 4/2/1955:14, WIFL; no author, "Smith's Bridge Loses Battle with Traffic, Ruled Unsafe," MN October 28, 1954:1, WIFL. Wilmer W. MacElree, *Along the Western Brandywine*, Second Edition (West Chester, PA: F. S. Hickman, 1912), p. 155. Five span lengths appear in the above sources: 126 feet, 136 feet 6 inches, 139 feet, 154 feet 6 inches, and 159 feet. The shorter lengths appear in the news articles; the longer ones in the Fitzsimons survey and on the Allen list. The

decrepancy may be due to the longer lengths incorporating the footage of the approaches. The 159 foot length comes from the Fitzsimons survey.

[24]Fitzsimons, *Historical Bridge Survey*, Smith's Bridge Survey; "Smith's Bridge Loses Battle," MN 10/28/1954:1, WIFL; "100 Year-Old Smith's Bridge," MN 12/17/1954:1, WIFL; MacDonald, "Historic Landmarks," JEE 4/2/1955, WIFL.

[25]Irving Warner letters to and from Henry F. du Pont, Maurice du Pont Lee, Richard A. Haber and Joe S. Robinson between June 27 and July 19, 1960, Irving Warner Collection, Accession 1518 Box 16, Hagley Museum and Library (hereafter HML); Virginia Delavas, "Neglect, vandals endangering our historic covered bridges," in MN August 23, 1974:41, WIFL; no author, "Covered Bridge Opens Again with New Lease on Life," JEE November 19, 1955:4, WIFL.

[26]Fitzsimons, *Historical Bridge Survey*, Thompson's Bridge Survey; Zebley, *Along the Brandywine*, p. 98; Bridge card file, Delaware Department of Transportation (hereafter DELDOT); 1959 Witheford bridge list, Gen. Ref. File 565, State Archives; William P. Frank, "Ancient Covered Bridges, like Covered Wagons, Pass in March of Progress," February 24, 1934 news article, William P. Frank Collection, Box 4, Historical Society of Delaware (hereafter HSD).

[27]1934-35 State Contract, Records of the Comptroller: Roads and Bridges contracts and Specifications, RF2450, State Archives; Letter, Leon deValinger to Miss Mildred Reynolds, April 1935, Gen. Ref. File 565, State Archives; Frank, "Ancient Covered Bridges," 2/24/1934, Frank Coll., Box 4, HSD.

[28]Zebley, *Along the Brandywine*, pp. 100-102; 1959 Witheford and Allen bridge lists, Gen. Ref. File 565, State Archives; William P. Frank, "Old Bridge Gives Way to New," JEE May 17, 1934, Frank Coll., Box 4, HSD.

[29]Zebley, *Along the Brandywine*, pp. 100-102; MacElree, *Western Brandywine*, p. 158.

[30]October 1934 State contract, Records of the Comptroller: Roads and Bridges contracts and specifications, RG2450, State Archives; MacElree, *Western Brandywine*, p. 158; Frank, "Ancient Bridges," 2/24/1934, Frank Coll., Box 4, HSD; Frank, "Old Bridge Gives Way," JEE 5/17/1934, Frank Coll, Box 4, HSD.

[31]Spero, *Historic Bridges*, pp. 46-47; Fitzsimons, *Historical Bridge Survey*, Rising Sun Bridge Survey: Zebley, *Along the Brandywine*, p. 150; *The Hagley Museum Guide*, (Wilmington, DE: Eleutherian Mills-Hagley Foundation, 1976), p. 12; Harold B. Hancock, "The Industrial Workman Along the Brandywine: 1800-1900," unpublished research report in three volumes, August 1958, Vol. 3, p. 112, Accession 1645 #35, HML; A. O. H. G., "A Tale of Two Old Covered Bridges," JEE May 12, 1945:6, WIFL.

[32]Spero, *Historic Bridges*, pp. 46-47; Fitzsimons, *Historical Bridge Survey*, Rising Sun Bridge Survey; Federal Writers Project, Vol. 16, p. 147, UofD.

[33]Hancock, "Industrial Workman, 1800-1840," Vol. 1, pp. 124 and "1870-1900," Vol. 3, pp. 110-111, 122, 158, Acc. 1645 #35, HML; MacElree, *Western Brandywine*, p. 180; Richard Sanders Allen, *Covered Bridges of the Middle Atlantic States* (Brattleboro, VT: The Stephen Greene Press, 1959), p. 41; Bryant, *Picturesque America*, pp. 223, 227; A. O. H. G., "A Tale of Two Old Covered Bridges," JEE 5/12/1945:6, WIFL.

[34]Bryant, *Picturesque America*, p. 227; Williams, *Illustrated History of the First State*, p. 62; Letter, Richard Sanders Allen to David K. Witheford, February 3, 1959, Gen. Ref. File 565, State Archives; 1959 Allen bridge list, Gen. Ref. File 565, State Archives; Federal Writers Project, Vol. 27, p. 48, UofD; Samuel C. Rumford, "The Brandywine Mills," Accession 601, HML.

[35]Spero, *Historic Bridges*, pp. 159-160; Fitzsimons, *Historical Bridge Survey*, North Market Street Bridge Survey; Scharf, *History of Delaware*, Vol. 2, pp. 670-671; Zebley, *Along the Brandywine*, pp. 178-179; Federal Writers Project, Vol. 16, p. 142, UofD;

A. O. H. G., "Bridges Spanning the Brandywine," JEE March 27, 1936:8, WIFL.

[36]Scharf, *History of Delaware,* Vol. 2, p. 670; Pursell, *That Never Failing Stream,* pp. 6-7, 45; Rumford, "The Brandywine Mills," Acc. 601, HML.

[37]Pursell, *That Never Failing Stream,* pp. 6-7; Federal Writers Project, Vol. 24, p. 48, UofD; Rumford, "The Brandywine Mills," Acc. 601, HML; Peter C. Welsh, "The Brandywine Mills: Chronicle of an Industry, 1762-1816," *Delaware History* (Wilmington, DE: Historical Society of Delaware, 1956), Vol. II, pp. 17-36.

[38]MacElree, *Western Brandywine,* pp. 192-193; Rumford, "The Brandywine Mills," Acc. 601, HML.

[39]Scharf, *History of Delaware,* Vol. 2, pp. 670-671; Welsh, "The Brandywine Mills," pp. 17-36; Zebley, *Along the Brandywine,* p. 179; A. O. H. G., "A Tale of Two Old Covered Bridges," JEE 5/12/1945:6, WIFL.

[40]Jamison, "Death and Life of Red Clay Creek," p. 5; Pursell, *That Never Failing Stream,* pp. vii, viii, 2-5; *Niles Register,* (Baltimore, MD: 1815), pp. 95-96.

[41]Federal Writers Project, Vol. 24, p. 90, UofD; Pursell, *That Never Failing Stream,* pp. 7, 30, 35, 105-110; 1959 Witheford and Allen bridge lists, Gen. Ref. File 565, State Archives.

[42]Pursell, *That Never Failing Stream,* pp. 37, 112-116; 1959 Witheford and Allen bridge lists, Gen. Ref. File 565, State Archives; Carroll Wirth Pursell, Jr., "The Delaware Iron Works," Delaware History (Wilmington, DE: Historical Society of Delaware, March 1959), Vol. VIII No. 3, p. 304; Carroll Wirth Pursell, Jr., *Ironworks on Red Clay Creek in the 19th Century* (Wilmington, DE: Historic Red Clay Valley, Inc., 1962), pp. 9-10, 15.'

[43]C. A. Weslager, *140 Years Along Old Public Road* (Wilmington,

DE: Historic Red Clay Valley, Inc., 1960), p. 22; September 1929 Contract, Records of the Comptroller: Roads and Bridges contracts and specifications, RG2450, State Archives; no author, "Passing of the Covered Bridge," SS June 16, 1929:12, WIFL.

[44]Pursell, *That Never Failing Stream*, pp. 34, 105-108; Jamison, "Death and Life of Red Clay Creek," p. 5; Weslager, *140 Years*, p. 15.

[45]"Passing of the Covered Bridge," SS 6/16/1929:12, WIFL.

[46]Spero, *Historic Bridges*, p. 26; Fitzsimons, *Historical Bridge Survey*, Ashland Bridge Survey; 1959 Witheford and Allen bridge lists, Gen. Ref. File 565, State Archives; "Ashland Bridge Planning Study," DELDOT, July 1979, Frank Coll., Box 4, HSD; Highway Council Report, November 9, 1974, Bridge File 1-118-258, DELDOT; National Register of Historic Places Supplement, 1974, Bridge File 1-118-258, DELDOT; Delavas, "Neglect, vandals...bridges," MN 8/23/1974, WIFL.

[47]Allen, *Covered Bridges*, pp. 42-43; Spero, *Historic Bridges*, pp. 26; Delavas, "Neglect, vandals...bridges," MN 8/23/1974, WIFL; *Hockessin Community News Town Crier*, "Ashland Covered Bridge no longer spans Red Clay Creek," November 10, 1982, Bridge File 1-118-258, DELDOT.

[48]Pursell, *That Never Failing Stream*, pp. 63, 109-110; Scharf, *History of Delaware*, Vol. 2, pp. 916, 924; "Ashland Bridge Planning Study," DELDOT, July 1979, Frank Coll., Box 4, HSD.

[49]Allen, *Covered Bridges*, p. 43; 1959 Witheford bridge list, Gen. Ref. File 565, State Archives; Delavas, "Neglect, vandals...bridges,"MN 8/23/1974, WIFL; Letters, J. W. Beretta to Henry B. du Pont, September 4 and 16, 1935, Henry B. du Pont Papers, Accession 1608 Box 20, HML.

[50]Henry B. du Pont letter of 1938 mentioning bridge destruction by flood, Acc. 1608 Box 110, HML; 1959 Witheford bridge list, Gen. Ref. File 565, State Archives.

⁵¹Spero, *Historic Bridges*, p. 110; Bridge File 1-120-261, DELDOT; 1922-1932 contracts regarding Bridge #120, Records of the Comptroller: Roads and Bridges contracts and specifications, RG2450, State Archives.

⁵²Pursell, *That Never Failing Stream*, pp. 111-112; Mt. Cuba Astronomical Observatory, *Proposed F. G. duPont Wing* (Greenville, DE: Mt. Cuba Astronomical Observatory, October 1972), p. 5.

⁵³Spero, *Historic Bridges*, p. 28; Fitzsimons, *Historical Bridge Survey*, Wooddale Bridge Survey; Bridge File 1-137-263A, DELDOT; 1959 Witheford and Allen bridge lists, Gen. Ref. File 565, State Archives.

⁵⁴Bridge File 1-137-263A, DELDOT; Delavas, "Neglect, vandals... bridges," MN 8/23/1974, WIFL.

⁵⁵Bridge File 1-137-263A, DELDOT; Pursell, *Ironworks*, p. 20.

⁵⁶Pursell, *Ironworks*, pp. 6, 7, 19-24; Pursell, *That Never Failing Stream*, pp. 37-38, 123-124; Pursell, "Delaware Iron Works," p. 304; Scharf, *History of Delaware*, Vol. 1, p. 398 and Vol. 2, p. 927.

⁵⁷Bridge File 1-003-225, DELDOT; 1959 Witheford bridge list, Gen. Ref. File 565, State Archives; Delaware State Highway Department 1920s photograph, HSD Bridge Photograph Collection.

⁵⁸Pursell, *That Never Failing Stream*, pp. 30-32, 126-128; Scharf, *History of Delaware*, p. 925; Francis A. Cooch, *Little Known History of Newark, Delaware and Its Environs* (Newark, DE: The Press of Kells, 1936), p. 106.

⁵⁹Pursell, *That Never Failing Stream*, p. 31; Weslager, *Forgotten River*, pp. 155-156.

⁶⁰Bridge File 001-147-270, DELDOT; Delaware State Highway Dept. 1920s photograph, HSD Bridge Photograph Collection.

⁶¹Pursell, *That Never Failing Stream*, pp. 30-31, 119; Weslager, *140 Years*, p. 15.

⁶²C. A. Weslager, *Brandywine Springs: The Rise and Fall of a Delaware Resort* (Wilmington, DE: Hambleton Company, Inc., 1949), pp. 7, 15, 59-68, 84, 96-97.

⁶³Weslager, *Forgotten River*, p. 155.

⁶⁴Scharf, *History of Delaware*, Vol. 2, p. 932; Weslager, *Forgotten River*, pp. 145-147.

⁶⁵Scharf, *History of Delaware*, Vol. 2, pp. 914, 927, 940, 942; Weslager, *Forgotten River*, pp. 147-149, 153-160; Cooch, *Newark, Delaware*, p. 115; Pursell, *That Never Failing Stream*, pp. 5, 8.

⁶⁶Weslager, *140 Years*, p. 153; Williams, *First State*, p.73.

⁶⁷Cooch, *Newark, Delaware*, p. 54; Scharf, *History of Delaware*, Vol. 2, pp. 916, 922-926, 937-939; Federal Writers Project, Vol. 24, pp. 82, 84, UofD; Constance J. Cooper, *The Curtis Paper Company: From Thomas Meeteer to the James River Corporation* (Wilmington, DE: The Cedar Tree Press, Inc., 1994), p. 3.

⁶⁸Cooper, *Curtis Paper Company*, p. 51; 1959 Witheford bridge list, Gen. Ref. File 565, State Archives; Delavas, "Neglect, vandals... bridges," MN 8/23/1974:41. WIFL.

⁶⁹Scharf, *History of Delaware*, Vol. 2, p. 922; Delaware State Highway Dept. 1925 photograph, HSD Bridge Photograph Collection; Daniel G. Beers, *Atlas of the State of Delaware* (Philadelphia, PA: Pomeroy & Beers, 1868), p. 19; G. William Baist. *Atlas of New Castle County, Delaware* (Philadelphia, PA: G. William Baist, 1893), Plate 10.

⁷⁰Spero, *Historic Bridges*, pp. 172-173; Scharf, *History of Delaware*, p. 933; Beers, *Atlas of the State of Delaware*, p. 19; Baist, *Atlas of New Castle County*, Plate 10.

⁷¹Cooper, *Curtis Paper Company*, p. 51; Records of the County Engineer, File CO-18, RG2460, State Archives; 1959 Witheford bridge list, Gen. Ref. File 565, State Archives; Bridge File for Bridge #231 and card file, DELDOT.

[72] Cooper, *Curtis Paper Company*, pp. 1, 3, 13-25, 52; Scharf, *History of Delaware*, pp. 922, 926.

[73] 1959 Witheford bridge list, Gen. Ref. File 565, State Archives, Delaware State Highway Dept. 1920s photograph, HSD Bridge Photograph Collection; Delaware Department of Transportation, Geographic Information Section, *General Highway Map—New Castle County* (Dover, DE: DELDOT, 1992).

[74] Scharf, *History of Delaware*, p. 939; Baist, *Atlas of New Castle County*, Plate 10.

[75] Spero, *Historic Bridges*, p. 24; Delavas, "Neglect, vandals...bridges," MN 8/23/1974:41, WIFL; Fred Hartmann, "Covered Bridge is Nation's Newest," MN November 21, 1960, WIFL.

[76] Cooch, *Newark, Delaware*, pp. 146, 160; Beers, *Atlas of the State of Delaware*, p. 19; Scharf, *History of Delaware*, p. 916.

[77] ASCE, *American Wooden Bridges*, pp. 3, 14-15, 47.

[78] Spero, *Historic Bridges*, p. 176; Scharf, *History of Delaware*, Vol. 1, p. 424.

[79] Spero, *Historic Bridges*, p. 176; Scharf, *History of Delaware*, Vol. 1, pp. 423-424; Allen, *Covered Bridges*, pp. 41-42; W. Emerson Wilson, "Woman Says Wrong Site Razed as Buck," MN October 1, 1963, WIFL.

[80] Scharf, *History of Delaware*, Vol. 2, p. 958; Wilson, "Woman Says Wrong Site," MN 10/1/1963, WIFL; 1959 Allen bridge list, Gen. Ref. File 565, State Archives.

[81] Hartmann, "Covered Bridge is Nation's Newest," MN 11/21/1960:1, WIFL; no author, "Newest Covered Bridge Has Some Enduring Features," *Every Evening* (hereafter EE), Wilmington, DE, November 21, 1960:25; Delavas, "Neglect, vandals...bridges," MN8/23/1974:41, WIFL.

[82] Hartmann, "Covered Bridge is Nation's Newest," MN 11/21/1960:1, WIFL; "Newest Covered Bridge," EE 11/21/1960:25;

National Society for the Preservation of Covered Bridges, *World Guide to Covered Bridges*, Rev. edition, Oscar F. Lane, ed. (South Peabody, Mass: National Society for the Preservation of Covered Bridges, Inc., 1972), p. 7.

[83]Lane, *World Guide*, p. 7; Delavas, "Neglect, vandals...bridges," MN 8/23/1974:41, WIFL.

[84]no author, "Another Covered Bridge Doomed by Levy Court," MN January 31, 1934:8, WIFL; Delavas, "Neglect, vandals...bridges," MN 8/23/1974, WIFL; Historic Red Clay Valley, Inc., *Preserving yesterday's heritage for tomorrow's future* (Wilmington, DE: Historic Red Clay Valley, Inc., [1976].

[85]List of Twentieth Century Paintings hanging in the DuPont Building and the Hotel duPont, November 6, 1959, Accession 1410 Box 57, DuPont Company Public Affairs files; list of paintings supplied by Richard C. Layton; no author, "Will soon be a Fragrant Memory," SS 8/12/1934:1, Magazine Section.

[86]"Brandywine Polka" sheet music, Terry A. Bryan, D.M.D., private collection of Delawariana; Brooks, "Covered Bridges Go," MN 8/24/1957, WIFL.

[87]Interview with Edward B. Cheney, June 1958, Accession 2026, HML; "Another Covered Bridge Doomed," MN 1/31/1934:8, WIFL.

Bridges

COLOPHON

    This book was composed in eleven point Goudy Oldstyle, two point leaded and printed on 80# Mohawk Superfine Text by The Cedar Tree Press, Inc., Wilmington, Delaware. It was bound by Advantage Bookbinding of Baltimore in cloth, using Curtis Tweedweave endleaves.

624
MCN

McNinch, Marjorie G.
Bridges

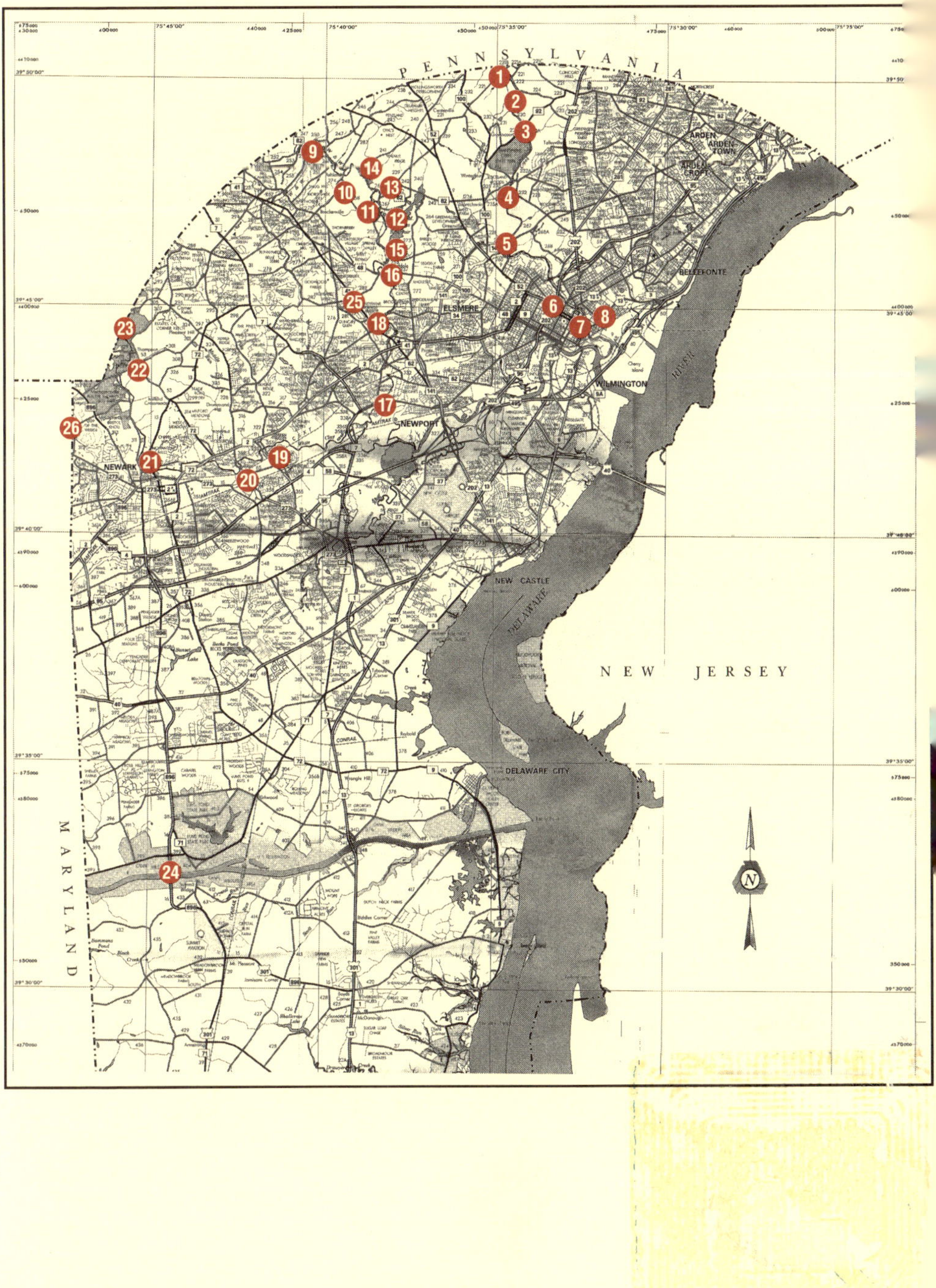